The Tactical Guide to Spiritual Warfare

The Tactical Guide to Spiritual Warfare

By Nicholas Anthony

Copyright 2023 by Nicholas Anthony (Nicholas DiRobbio)

All rights reserved. No part of this document may be reproduced or transmitted in any form or by any means, electronic, mechanical, photocopying, recording, or otherwise, without prior written permission by Nicholas Anthony. Requests for permission to make copies of any part of the work should be submitted to the publisher.

Nicholas Anthony
www.officernicholasanthony.com
Invisawounds34@gmail.com
ISBN(Print) 978-1-7351561-2-5

Unless otherwise indicated, all Scripture quotations are taken from the *Holy Bible*, New Living Translation, copyright © 1996, 2004, 2015 by Tyndale House Foundation. Used by permission of Tyndale House Publishers, Carol Stream, Illinois 60188. All rights reserved.

"Scripture quotations marked (ESV) are from The ESV® Bible (The Holy Bible, English Standard Version®), copyright © 2001 by Crossway, a publishing ministry of Good News Publishers. Used by permission. All rights reserved."

Scripture quotations marked (NIV) are taken from the Holy Bible, New International Version®, NIV®. Copyright © 1973, 1978, 1984, 2011 by Biblica, Inc.™ Used by permission of Zondervan. All rights reserved worldwide. www.zondervan.comThe "NIV" and "New International Version" are trademarks registered in the United States Patent and Trademark Office by Biblica, Inc.™

Scripture quotations marked MSG are taken from *The Message*, copyright © 1993, 2002, 2018 by Eugene H. Peterson. Used by permission of NavPress. All rights reserved. Represented by Tyndale House Publishers.

Scripture quotations marked (CEV) are from the Contemporary English Version Copyright © 1991, 1992, 1995 by American Bible Society, Used by Permission. Scripture quotations marked KJV are taken from the King James Version (Public Domain).

Dedication

To my daughters, Aliana Rose, and Daniella Marie.

May you both become warriors in Christ.

> "Praise the LORD, who is my rock. He trains my hands for war and gives my fingers skill for battle."
>
> Psalm 144:1 NLT

Table of Contents

Introduction i
1. The Battlefield 1
2. The Armor Of God 9
3. Night Operations 19
4. Know The Enemy 29
5. Principles of Security 37
6. Defensive Operations 49
7. Patrolling The Battlefield 57
8. Three-Dimensional Warfare 67
9. Friendly Fire 77
10. Warrior Leadership 87
11. Battle Fatigue 95
12. Offensive Operations 105
About the Author 112

Introduction

The United States Army issues several books of tactical doctrine known collectively as field manuals or "FM's" for short. Their subjects deal with anything from planning and operations to first aid, psychological operations, small unit tactics, and more. The Army has field manuals for just about everything you can think of. They would even have one about going to the bathroom if they thought it necessary and guess what...they did and they do! (FM 21-10 Field Hygiene and Sanitation).

Perhaps one of the more prolific field manuals known to Army Officers is FM 7-8, Infantry Rifle Platoon and Squad. Any Officer Candidate, West Point Cadet, or ROTC Cadet can tell you the significance of this field manual for an Officer in training.

As an Army Reserve Officer Training Corps Cadet flying to Ft. Lewis in Washington State for Leadership Development Assessment Camp (LDAC, or "Advanced Camp") I remember seeing other Cadets in the airport terminal at Seattle-Tacoma International with their heads buried in a copy of FM 7-8, preparing for what they would soon be tested on in a field environment.

FM 7-8 is the bread and butter of small-unit Infantry tactics. As Platoon Leaders, Army Officers regardless of their final branch assignment, need to be well-versed in basic infantry tactics. These tactics are practiced and studied throughout a cadet's journey and put to test at advanced camp.

As I write this, 17 years removed from my Cadet days and 11 years removed from the Army, I can still recall the concepts and tactics. I feel confident I could still lead a platoon or squad in battle drills.

It wasn't until recently though that I figured out why I was being reminded about these tactics and I began to apply them to a new way of fighting.

To start, I will give you a brief history of my background which is talked about in length in my memoir, *Invisible Wounds, A Cop's Journey of Faith through the Darkness of PTSD.*

I had always wanted to be something special. As a kid, I loved all things Special Forces and the military. Once in college, I joined the Army ROTC program as a freshman. In my sophomore year, I also joined the Army National Guard under a dual-membership program with ROTC, which allowed me to drill both with an Army National Guard unit and my ROTC Battalion.

In 2007, I was commissioned as Army 2nd Lieutenant and served part-time as a platoon leader and later as a company executive officer in an Army National Guard, Military Police Company. After graduating college that summer, I was hired as a Police Officer and attended the state's Police Training Academy. For the next four years, I wore two hats as a full-time police officer and part-time Army Lieutenant.

Here's where God came in. I was suddenly removed from the Army in 2011 for a rather bizarre reason, a food allergy, which they were already aware of when I joined in 2005. Nevertheless, God's hand was at work then for unknown reasons and would later be at work again at the 12th year mark of my Law Enforcement career. Detailed in *Invisible Wounds*, I developed what is called Cumulative Post Traumatic Stress Disorder or CPTSD and suffered from severe anxiety and depression.

After a two-year battle with the municipality I worked for, I retired at 14 years with no recognition or benefits entitled to me under what should have been a work-related injury. Suffice it to say, it has been a real challenge during my faith journey to accept and understand why things played out the way they did. Then again, the Bible says that God's ways are not our ways (Isaiah 55:8) so who are we to question the Almighty? I suppose everything will be explained in great detail when Jesus and I have our face-to-face someday in heaven.

During my struggle with PTSD, I also struggled with God and faith. I turned to the Bible, looked inside myself, and saw that my problems went far behind PTSD from Law Enforcement and included things such as lust and pride.

In the years after "retirement," if you want to call it that, I discovered I had an anointing for faith-based writing. Not only did I write, *Invisible Wounds,* a faith-based book, but I also wrote several prophetic journal entries and started an online devotional ministry, and well here I am typing this. And truth be told, I have come under attack even as I started this project.

During my devotional writings, God began to teach me hidden connections and meanings within the scriptures that I never saw before. I began to see Bible stories differently and find importance in scripture that I had previously discarded or skipped over.

I saw Old Testament stories and revelations that pointed to Jesus in a new way by connecting spiritual dots in the scriptures. I would write a devotional and the next day, I would find myself going back to that same devotional to apply the very lesson I wrote about. One can only attribute this to the Holy Spirit speaking and teaching me directly through my own writing.

I began to notice the Holy Spirit pointing me toward my past experience and passion for military and law enforcement tactics. I started applying them in my writings to the scripture and warfare in the spiritual realm.

When I was working in Law Enforcement, I had a passion for tactics and teaching. I was certified as an instructor in all kinds of things from firearms to tactical medicine, and active shooter/threat tactics. As a Platoon Leader in the National Guard, teaching tactics and running firearms ranges was also something that I often did.

I write this as someone who has fought and continues to fight the spiritual battle as a Christian. I realized that although I am no longer fulfilling the role of Police Officer/ Soldier in the physical sense, my past life experience and training were preparing me to apply the warrior concepts I had previously learned to a new kind of battle, a spiritual one. This was a battle that I needed to be stronger in if I were to perform and fulfill my kingdom assignment.

With that said, whether or not you have ever served in the capacity of a warrior, I want you to start viewing yourself as one. Every single Christian is called to be a warrior. If you profess to be a Christian and think you are going to sit peacefully and idly by, outside of the fight, then you are sadly mistaken.

It should be no surprise to all that we are closing in on end times. If you are staring up at the sky waiting for the rapture or burying your head in the sand until Jesus returns then you are not fulfilling your calling from God during this time. Just like the angels that appeared to the disciples in Acts 1:11 NIV, you are being told to get to work in Jesus's absence:

"Men of Galilee," they said, "why do you stand here looking into the sky? This same Jesus, who has been taken from you into heaven, will come back in the same way you have seen him go into heaven."

What this book is going to do is give you tried and true battle-proven concepts and tactics. These tactics not only apply to the physical battlefield but the spiritual one. Let's be clear, this book is from God and about God. Though I am using military and tactical analogies, don't lose sight of the overarching picture which is warring in the spirit.

I am going to show you how the warrior mindset and living a warrior life translates to your walk with God.

I believe God is using my experience and passion in this area to apply it to a new battle...what I am going to tell you is the most important battle you and I will ever face. Remember this guide is not here to make you an infantry soldier, it is meant to make you a warrior in the Army of God.

Father God, before we start, I want to invite You into this book Lord. I want to invite the Holy Spirit in every letter, word, and page. I want Your spirit to guide its message to the reader, to convey Your heart. I want its message to point people toward Jesus through its pages.

I ask that You guide my hands and strengthen my mind and send protection to both the reader and myself as they take on the darkness. I pray that my flesh and pride are absent and apart from your will. To include the success and purpose of this book. I asked that many are blessed and strengthened through it and inspired to live under your command. Anoint this book with overflowing oil and the blood of your son Yeshua, in Jesus's mighty name, Amen.

1.
The Battlefield

A battlefield is where two or more forces come to engage in combat. That's possibly the simplest explanation I can give it. When we think of a battlefield, we think of a field, with barbed wire, trenches, and craters formed from explosives. You may envision confusion, chaos, fire, and smoke. While all these things may be correct, the battlefield has changed over the years even in the physical realm.

For example, The American Revolutionary and Civil Wars were fought out in the open, firing line to firing line. WWI was fought in trenches across a "no man's land", WWII was fought in European cities, towns, beaches, and the African desert. Vietnam was mainly fought in the jungle. Iraq was fought in cities, deserts, and wheat fields, and Afghanistan in the mountains.

The terrain may change, but the concept remains the same... warfare and conflict. Since the beginning of the fall of man (Adam and Eve in The Garden) man has engaged in an unseen war...a spiritual one.

> The scripture says, "For we wrestle not against flesh and blood, but against principalities, against powers, against the rulers of the darkness of this world, against spiritual wickedness in high places." Ephesians 6:12 KJV.

We "wrestle" or as the Holeman Christian Standard Bible says "we battle". That is we contest against it, we war with it.

War, both physical and spiritual is part of our fallen world. Wishing for world peace is a nice concept, but it is not biblical or realistic. Until the return of Jesus Christ and the new heaven and earth, there will continue to be physical battles and spiritual conflicts.

The World vs. The Word

The events of 2019 to the present time brought up this conflict to the forefront with Covid, abortion, and other hot political issues. Instead of seeing a united front in the Church under the truth of the Bible, we saw church pastors closing doors, and Christians supporting woke ideology, and anti-biblical principles. Fear had gripped the public and the body of Christ, who became even more fractured by masks, and treatment options.

Individuals and families turned against one another. And people put the government agencies and medical "experts" or media-controlled talking heads higher than God in their position of trust. Roe V Wade was overturned by the Supreme Court, and suddenly you saw where certain Christians and churches stood, instead of the biblical principles they should have stood on.

But this book isn't about those issues, as they should not be controversial for you as a Christian. If you don't know where to stand yet on those issues, then put this book down and go read the best guide of them all...The Holy Bible first.

I say this because you cannot begin to be a warrior for God in the spiritual realm if you are still confused and tied to the ways of the world.

> Romans 12:2 KJV says, "And be not conformed to this world: but be ye transformed by the renewing of your mind, that ye may prove what is that good, and acceptable, and perfect, will of God."

That means we don't follow what the world is doing, instead we seek the will and heart of God so we know what is acceptable to him. This should be a simple concept, or at least it should be to a Christian, but so many people chose to live first in the world, versus out of the world. The problem is that so many people want to be accepted if not loved by the world. They don't have core beliefs grounded deep enough in their own faith not to be swayed by what is "trending" on social media.

We saw a perfect example of this with the war on cops. Many people I was personal friends with who probably never had a negative interaction with a cop, were suddenly anti-law enforcement. Had the trend been against say... postal workers, then everyone would hate postal workers. No matter how ridiculous this is, it appears people no longer have independent minds to think for themselves.

This is why the scripture says that you must first remove yourself from the world.

> John 15:19 NLT says, "The world would love you as one of its own if you belonged to it, but you are no longer part of the world. I chose you to come out of the world, so it hates you."

Of course, the world would love you as its own if you embrace it and succumb to it. But you are Christian, a warrior, and your commander-in-chief is God, not man.

As Peter and the other Apostles said in Acts 5:29 NLT *"We must obey God rather than any human authority."* That is we don't live for this world, we don't love things of this world, and we set ourselves above the things of this world. (1 John 2:15, Colossians 3:2)

Only then, once we have separated ourselves from this world can we recognize what is deception and the doctrines of demons among the world *and* the church (1 Timothy 4:1).

Among the church you say? Yes, the same scripture in 1 Timothy 4:1 warns us that these things will happen in later times as people fall from what is biblical faith.

This is echoed again in 2 Timothy 4:3 ESV,

> "For the time is coming when people will not endure sound teaching, but having itching ears they will accumulate for themselves teachers to suit their own passions."

Translation, "I don't like what the Bible is saying, so I will find a church or pastor who agrees with my sin and my lifestyle."

Has this happened? It sure has, and if your church or pastor is supporting anything that is against biblical marriage, life, or sin, you should run...very far away from that church. My friends, don't even be surprised if the self-proclaimed "supreme authority" on scripture, the Catholic Church, begins to walk back on biblical truth in later times either.

This you can take to heart though, there is only one true church and denomination and that is the church of Jesus Christ that we see in the book of Acts. All ministries and every Christian should be emulating the acts of the apostles and the great commission Jesus gave to all believers. The truth is we are "The Church".

The Bible says that we are a "kingdom of priests", a "royal priesthood" and a "holy nation" sent out of darkness to preach the light (Isaiah 61:6, Exodus 19:6 1 Peter 2:9, Revelation 1:6 ESV).

Do you realize that is said in four separate scriptures? The books of Isaiah and Revelation were written several hundreds of years apart, but the message remains the same!

God is telling us that we are meant to be priests, or as Saint Teresa put it, "the hands and feet" of Jesus.

So, before you can set foot on the battlefield, you must first understand the conflict of "the world vs the word." You must understand that the battlefield is set up in such a way that it doesn't look like one. It is said that one of the greatest tricks of the enemy (Satan) is to convince people that he doesn't exist.

I would quote who said this but it's been accredited to multiple people throughout the ages.

Lines of Communication

The enemy has gotten us used to a life on the battlefield, that we no longer even see it anymore.

You are going to need spiritual eyes to be able to fight, and that requires you constantly seek a connection to God and the Holy Spirit.

To illustrate this point, even in a military operation there is always a connection, a communication line to the command center. The command center is like the eye of the battlefield from above. It is monitoring the battle from a viewpoint that you as a ground soldier are unable to see.

You will have to maintain your "comms" and radio frequency tuned into the Holy Spirit. The Holy Spirit is the command center that keeps you focused on your objective amidst the fog of the battlefield. The Holy Spirit is like a frequency over radio waves and finding it requires spiritually tuning into the right station.

Because there is such chaos at war, it is definitely possible to pick up the wrong frequency while on the battlefield. The enemy could use this frequency to mislead you or send you the wrong information and confuse you. This is called psychological warfare operations. We will discuss this topic in greater detail later in the guide.

Please understand that discerning lines of communication are important as a Christian warrior.

Macro vs. Micro Warfare

We talked about what a battlefield is, we talked about recognizing the battlefield, and now let's discuss some of the different places conflicts take place.

Ephesians 6:12 talks about the "principalities of darkness." A principality is like a region, state, or territory. It's a large area, ruled by the enemy. A principality of darkness could be among a school system, congressional body, the media, or even a presidential administration, or royal kingdom.

The Old Testament is full of examples of evil kingdoms and rulers. Though they may have changed names over the centuries, the principalities behind them remain the same.

They control and move chess pieces strategically placing their forces on the battlefield. They always seek out weaknesses and ways to exploit facets of society to influence large groups of people, down to the individual.

Now that leads us to the micro battlefield, perhaps the most important reason why you are here reading this, the battlefield of the mind.

You see our minds are an organ--flesh, and blood. And just like all flesh, they are fallen and subject to corruption, decay, disease, and death. Newsflash, when you die, your mind dies, so how then do you live on with a conscious in the afterlife?

That is a great question, it's because you have a spirit man inside of you who houses your soul. This, my friends, is actually who you are. Your soul was breathed to life by God and knit together with your body in your mother's womb (Psalm 139:13) The NIV version says *"you created my inmost being."* Your soul is your inmost being, it was created by God, and it wants to return to God when your flesh dies.

So the battle is not just for the mind but also the soul. You can love Jesus in your heart but be corrupted in the mind. But how? Well, here are just a few examples I have struggled with myself...PTSD, depression, obsessive-compulsive intruding thoughts, and even lust.

I believe in Jesus, and I accept Jesus as my Lord and savior, but my mind is flesh. This is why Jesus says in...

> Matthew 26:41 NIV, *"Watch and pray so that you will not fall into temptation. The spirit is willing, but the flesh is weak."*

Yes, the flesh is indeed weak my friends, and something we as believers always have to contend with. There is a saying that "where the mind goes, the body will follow."

The Bible more accurately says:

> "For as he thinketh in his heart, so is he" Proverbs 23:7 KJV.

And the original Hebrew word used for "heart" in this scripture is *"nephesh"* which means soul. So your soul is what matters, but most of the battle occurs in your flesh, in the mind.

Your mind is a place where radio frequencies of the spirit realm both good and bad are intercepted. Some, like the prophetic, are more affected and in tune with these frequencies which can be a blessing and a curse. Yours truly has continued to fight this battle, and it even tried to stop me from writing this book. Just prior to being this project, I came under a wave of attack. This is usually an indicator that you are onto something.

But we have to remember in these moments of attack that God didn't give us a spirit of fear but of power, love, and soundness of mind (2 Timothy 1:7).

The mind is often influenced by external factors, such as trauma, abuse, drugs, and even pornography. All these things affect the mind on a physiological and spiritual level. It is true that most mental health issues arise from some kind of external stimulus like the ones listed above.

Scientists have argued that someone might be genetically predisposed to mental health issues but it's hard to believe that would be by the design of God. Perhaps a good example is how among first responders, and soldiers, someone can experience trauma and not be affected while others may develop PTSD from the same exposure.

This still largely remains a mystery to doctors, but I would argue from experience that a large percentage of this is spiritual. Unfortunately, whether spiritual or not, the mind can be actually "injured" by trauma.

But God can heal anything if He so chooses, and He has allowed this specific thorn in my flesh. In 2 Corinthians 2:12, the apostle Paul speaks of a similar "thorn in the flesh" from the enemy that the Lord has not removed. Paul never tells us what that thorn is but he says the Lord has kept it there to keep him from becoming prideful.

Perhaps I would not be here writing about this very topic of spiritual warfare if I had not had to endure it myself. At the end of the day, I have faith God will bring everlasting comfort so that I too may bring comfort and victory to others. (2 Corinthians 1:3-4)

The "struggle is real" my friends. There is a battle going on for the mind, bodies, and souls of believers. That's why we are called to equip ourselves with the "Armor of God" mentioned in Ephesians chapter 6. Just like a soldier in battle, the next chapter will talk about what kind of equipment you must take into the fight and how to maintain and even repair it! So get ready to put on the Armor of God.

2.

The Armor of God

This subject is really a "no-brainer" when it comes to military analogies in the Bible, but it's nevertheless important in your defense. You wouldn't step onto a battlefield without armor, equipment, or a weapon. Ok, David slayed Goliath without armor and just a measly slingshot but he had the spiritual armor and weapons of God. For the sake of military analogy, we are going to talk about physical armor as well.

From the days of breastplates, shields, and chain mail to the modern soldier, armor always served the same purpose, to deflect against attacks. Often times the purpose of armor is to protect our vital organs. For us, this means our hearts and minds!

Let's talk about what the scriptures say about attacks, weapons, and armor in this chapter, along with how to use and maintain them. Equipment needs to be both practiced with and maintained. or both your skill and the equipment will degrade over time.

Sure, weapons and armor have changed since Bible days, but the principle and purpose of armor remain the same; to protect the fighter on the battlefield.

The modern soldier has a kevlar helmet, and heavy-duty ceramic plates to protect the abdomen and back. Even the modern police officer wears body armor to include soft and hard ballistic plates.

As a police officer, I wore body armor every day of my career. It was just part of the uniform. I wouldn't dream of going out on the road without it. I owed it to my family and myself to return home at the end of the shift.

Being extra cautious, I even added a special threat trauma plate in my vest that I purchased with my own money for eighty dollars, something I considered a wise investment. I also carried a ballistic helmet and tactical plate carrier with heavier-duty ceramic hard plates for any active shooter events.

Even as a police officer on the streets, I had to be prepared to battle evil. And let me tell you, there was a lot of it out there. There is just so much sickness, hatred, and malice out there, that I often felt I was on the front lines of the end of the world. Again, the spiritual battlefield is everywhere.

This is why modern cops are increasingly looking more and more like soldiers. Unfortunately, evil is not decreasing in society but just the opposite. It's becoming more and more like the zombie apocalypse out there as humanity seems to be continually growing colder and spiraling into darkness.

Police have learned to adapt their tactics and equipment over the years and believe me for most law enforcement agencies this is not an easy task because they tend to be resistant to change. Law enforcement agencies are also notoriously cheap due to low budgets for training and equipment.

When I first got in the department for example we had shotguns in the cruiser, years later, we had "upgraded" to M4 rifles, but to prove my previous point, our shotguns were over twenty years old and our rifles 15 years old. Equipment and training should be much more of a priority in law enforcement agencies but sadly that is just not the case.

The modern cop should be carrying a personal radio, duty belt, first aid kit, body camera, Conducted Electrical Weapon, handgun, flashlights, spare ammunition, handcuffs, OC spray, and baton. I say "should" be carrying because not all departments are created equal. I was personally surprised to learn, some agencies were still operating without computers in the patrol car and handwriting tickets.

Whether you are a cop or soldier, your weapon and gear are important. You need to set them up to where you can easily access them, and you have to make sure they are clean and in proper working order. Most importantly you have to train with them!

In the same way, your spiritual weapons of war have to be trained. A good gunfighter can feel when his weapon runs empty. He knows the feeling and sound of the slide of a pistol or bolt of a rifle locking back on an empty weapon. This sound and feeling immediately trigger a subconscious action from the shooter to reload because it's been trained in them with repetition and muscle memory. But getting to this level takes practice.

When I went to law enforcement firearms Instructor school at the Smith &Wesson Academy, we shot over two thousand rounds each in a matter of two weeks. Most police officers will never get that amount of practice with their firearms in their lifetime unless they are in a specialty unit or spend their own time and money.

In the same way, the scripture is your weapon, and it needs to be there at the draw just like your firearm or a sword, at your side and ready to be drawn when a threat comes. The good news is that reading the Bible is free and doesn't cost you anything other than your dedication and time.

> 2 Corinthians 10:4 NIV says, "The weapons we fight with are not the weapons of the world. On the contrary, they have divine power to demolish strongholds."

The NLT versions (v3-4) say:

> "We are human, but we don't wage war as humans do. We use God's mighty weapons, not worldly weapons…"

So, you're probably thinking I just biblical disputed my own book right? Actually, reader, I am stressing the point here that again, this book is about *spiritual* weapons and tactics.

Clearly, bullets do nothing against spiritual forces. But I believe God wants us to make the analogy because the scripture gives us the pieces of spiritual battle armor listed out in Ephesians chapter 6. My online devotional ministry on Substack, *The Ephesians 6:11 Armory* is named after this very chapter in Ephesians.

I call it an "armory" because an armory houses weapons and as a spiritual armorer, I distribute them to people. I am assigned to give you the weapons that you will use to take part in the spiritual war. Now let's take a closer look at the items listed in Ephesians chapter 6 and their purpose and meaning.

> The Apostle Paul says "Put on **all of God's armor** so that you will be able to stand firm against **all strategies** of the devil." Ephesians 6:11 NLT (emphasis added).

There are two key points that I put in bold. You need **all** of the armor to withstand **all** of the attacks. I am not going to walk out on a battlefield or street as a cop missing any of my equipment.

It also does me no good to have a weapon, but no ammunition or a weapon but no armor. I was one of only a handful of cops in my department that carried additional ammunition and armor when out on patrol. Unfortunately, some "warriors"(and I use that term lightly) never think they are going to come under attack. But you know better and will be prepared!

The Belt and The Breastplate

It is important to realize that the warrior carries both defensive and offensive equipment.

The Armor of God | 13

> Verse 14 of the NLT says, "Stand your ground, putting on the belt of truth and the body armor of God's righteousness."

This is a great cop reference right here. The duty belt is the centerpiece of the police officer's gear. It is also the reason why my back hurts! It is said that the average police officer's duty belt weighs thirty pounds! Try chasing a suspect with that on, not easy.

What does a belt do, it circles around us (holds our pants up)...and well...a police belt holds tools and your weapon. The weapons in this case in the spiritual world, are the truth.

Listen, the enemy is a liar, and Jesus calls him the father of lies (John 8:44). Lies are one of his main weapons because that is really all he has. But the truth is like a myriad of tools that surround you like a belt.

This is really the key right here, we could end the book right here on this fact...but there is soo much more to talk about! Now notice in the same scripture, Paul says the belt of truth **and** the body armor of righteousness.

Listen to me...this is a big **AND** to pay attention to. Remember I said the scripture said to put on **all** of the armor. This scripture just doesn't say body armor...it says the body armor of righteousness. What is righteousness? It is living a certain way, following God's commandments, being obedient, and living right in his eyes.

Yes, this is a constant "battle" as well...pun intended. Living as a righteous person in a fallen world, in a fallen body is no doubt not easy. But righteousness is required for the armor to function properly. If there are self-imposed chinks in your armor, then you leave room for the enemy's arrows to pass through. So it is very important to try to live as Christ, though we know we will always fall short.

> Romans 3:23 KJV says "For we all have sinned, and come short of the glory of God.

But that doesn't mean we don't try. We have to strive for righteousness.

> Paul says in his letter to Timothy "But you, Timothy, are a man of God; so run from all these evil things. **Pursue righteousness** and a godly life, along with faith, love, perseverance, and gentleness." 1 Timothy 6:11 NLT(emphasis added).

Did you know body armor expires? Police officers change their vest every three to five years because the kevlar fabric degrades over time. Between the humidity and gallons of sweat that get soaked up through the years, the material loses its effectiveness. Even hard ceramic plates have to be x rayed and examined after a while to determine their usability.

Ok so you can't replace your spiritual armor every couple of years, but you can renew it and refresh your righteousness, by re-examining your faith and works. And you need to have both for a functioning breastplate of armor. James 2:17 says faith without works is dead. So living righteous and being obedient requires not just believing in Jesus, but a call to action...or works.

Jesus gives us many calls to action:

- Keep his commands (John 14:15)
- Feed his lambs (John 21:15)
- Forgive (Matthew 18:21-22)
- Preach (Matthew 28:19-20)

These are just some examples that are considered "works" to accompany your faith for righteousness.

The Shoes of Peace

> "For shoes, put on the peace that comes from the Good News so that you will be fully prepared." Ephesians 6:15 NLT

Footwear is important for both police and soldiers. Boots are the common choice for both professions. Usually, something steel-toe with a good amount of traction is used. This traction is needed to ground you.

Ground you to what? Ground you with the peace God has to offer when you face battles. I never understood the shoes of peace until recently. Some may take it as to bring peace to others, but instead, it is to ground you in comfort...in peace in battle and among the enemy.

Where does that peace come from? It comes from "the good news". What is the "good news?" The good news is that no matter what you are going through, the end of the story has already been written. Jesus died for your sin, and your salvation. The war has already been won!

Spoiler alert, God wins in the end. God knows the outcome, the beginning from the end, and the end from the beginning (Isaiah 46:10).

This is something our human minds cannot truly comprehend. It's like trying to wrap your head around the plot of a time-travel movie. It just gets frustrating trying to figure it out so don't try. I have to come to realize that we aren't meant to understand all the ways of God.

So, as you fight, lace up your boots and remember the promise of Jesus.

> "And the peace of God, which transcends all understanding" Philippians 4:7 NIV

The Shield, The Helmet, and The Sword

> "In addition to all of these, hold up the shield of faith to stop the fiery arrows of the devil" Ephesians 6:16 NLT

Cops can deploy ballistic shields in certain scenarios. It's common to have a shield at the front of the entry team in high-risk patrol or SWAT operations. Your shield is believing in God's promises.

> Psalm 91:4 NLT says *"His faithful promises are your armor and protection.*

Psalm 91 verses 5 and 6 says that if you have faith, then you don't need to fear the terrors by night nor the arrows by day. Not even deadly diseases should stand in your way.

So, when you hold up your shield of faith imagine that it says "PSALM 91" on it in big bold letters. It is a declaration and promise of protection you have the authority to carry and call upon.

Lastly, the helmet of salvation and the sword.

> "Put on salvation as your helmet, and take the sword of the Spirit, which is the word of God." Ephesians 6: 17 NLT.

A helmet protects your dome which is your head. Soldiers wear helmets that aren't just for ballistic protection but also for protection from falling debris and rubble.

In war, there will be explosions, and incoming indirect fire(mortars). Even if you aren't hit directly, it is very likely that pieces of building debris or shrapnel can fly your way.

The enemy bombards us with intrusive thoughts, negative thoughts, and worry. We will discuss this more in detail in other chapters, but your helmet is one of salvation, regardless of where your mind goes with your thoughts, if your heart is in Jesus then you have salvation.

The last piece of equipment is the sword or word of God. While armor and shields are more defensive in nature, a weapon can be used for both defense and offense.

A sword can stop a strike or give a strike. The same applies to scripture when facing incoming blows and delivering blows. The scripture is designed to be a weapon of words to stop attacks and deliver them to dark forces.

Now let's get modernized in our weapons for a minute. Every Soldier and Marine is issued a weapon, nowadays, it's likely an M4 or carbine version of the original M16. I've used both in my time in the military.

Some soldiers are issued heavier weapons like belt-felt machine guns and some are given designated marksmen weapons. Either way, each has a purpose. One delivers pin-point long-range accuracy and one delivers a large amount of suppressive fire.

As a warrior, you should be well-versed in both. You need to have a solid general understanding and overview of the Bible, but on occasion, you need to deliver the pin-point accuracy of scripture.

Soldiers and Marines know their weapons inside and out. They can assemble and disassemble them blindfolded; they even sleep next to them. You have to be the same way with the word of God. It is your weapon; you need to know it inside and out. You also need to be able to deploy it when you need to engage the enemy and do so proficiently.

As a police firearms instructor, I can tell you firsthand, not every cop practiced and was proficient with his or her weapon as I mentioned before. Unfortunately, the same thing applies to soldiers...not every soldier was a dead eye.

I once spent two whole range days with a female Army Major just trying to get her to hit the paper at 25 yards, let alone the target. Shooting and weapons manipulation are perishable skills, they must be practiced and become muscle memory. In the same way, you must do that with the word of God.

No one expects you to be able to recite scriptures and school people like Jesus did to the Pharisees, but you should have a solid foundation of scripture in your armory. There are some key scriptures of battle, many of which are mentioned in this chapter, that you should specifically commit to memory. Have them holstered like a pistol ready to be presented when a threat comes.

In the next chapter, we're going to talk about another piece of vital equipment...actually two pieces.

Lights and Night Vision. Both will allow you to see through the darkness.

3.
Night Operations

A battlefield is a dark and scary place at night. The one who "sees" in the dark has the advantage. "We own the night" used to be a slogan of the U.S Military. During Operation Desert Storm in the early 1990s, night vision capabilities gave us tremendous advantages over the Iraqi Armed Forces. I say "used to be" because that edge we had tactically as a fighting force at night doesn't really exist anymore. This is because most militaries now have night vision capability.

One should never underestimate the ability to see in the dark. Light is a powerful tool for the Christian believer. Light can illuminate a battlefield, but it can also give away your position to the enemy. You see up until this point, your sin kept you hidden from enemy forces. You weren't a problem or a noticeable threat to the enemy. But now that you shine the light of God, you will draw attention to yourself as an adversary, an enemy combatant, and you have given away your position. This isn't a bad thing to bear the light of God and we are going to talk about why in this chapter.

Light For Identification

Soldiers and cops for that matter are taught what is called "noise and light" discipline. In the military, red light is often used to see at night, because it doesn't ruin your natural night vision. It is also not as visible from far off to the enemy. Light penetrates the darkness, but again, this can be good and bad. Just a small light can be seen from far away in the dark. Researchers have put this to the test and believe that the human eye can detect a candlelight flame up to 1.6 miles away.[1] Surely the smallest amount of light could draw attention to yourself, and as a believer, you are going to be full of light. But your light is also a weapon and scripture says it is actually not to be concealed. But for the moment we are going to talk about light for the purpose of seeing and not being seen.

Soldiers use night vision headsets as one method to see in the dark without being seen. Night vision works by taking existing infrared light in the environment and magnifying it, creating that notorious green glow in a soldier's headset. Soldiers also have devices that can be mounted on rifles, and beacons that emit infrared light, which is invisible to the naked eye.

Infrared light can also be projected from soldiers wearing identification markers on their uniforms that emit signals of IR light to friendly aircraft or other forces in the area. Just like the invisible spiritual light that one believer may see in another believer, the military uses a similar concept to identify friendly forces on the battlefield with light only they can see. We as Christian warriors should carry our own IR light beacons and identifiers by our actions and internal light so we are recognized by other believers.

Soldiers can also use IR light to project an invisible beam that works like a flashlight that only they can see through their night vision device. The concept is actually pretty cool, and during Military Operations and Urban Terrain (MOUT) exercise, I got to experience using night vision and IR illumination while conducting night training raids on mock city buildings.

---

1 Krisciunas, K., & Carona, D. (2015). At What Distance Can the Human Eye Detect a Candle Flame? *ArXiv*. https://doi.org/10.48550/arXiv.1507.06270

Light For Searching

Now as a police officer, working nights, your flashlight is probably the most important tool on your belt. When I was training with my FTO or field training officer, he told me a principle regarding flashlights I will never forget. He said to me "two (flashlights) are one, one is none."

What does that mean? It means that things break, batteries die, and to put it in simpler terms, you need to carry multiple flashlights. You don't want to be caught on night shift with no flashlight, or a dead one. I used to carry two on my duty belt when I worked the night shift--one primary rechargeable and one smaller battery-powered one. Later on down the road, we were also issued tactical flashlights for our duty handguns as well, so technically three lights.

For a cop, the primary purpose of a light is for exposing threats and searching the darkness. As a spiritual warrior, you will also do the same.

It wasn't uncommon to search for suspects hiding in the woods, buildings, or behind houses. What does a cop do at a car stop at night? He lights the vehicle up with his spotlight and approaches with his flashlight. He is checking your vehicle for threats and contraband. In the same way, you will use your spiritual light to search the darkness.

Soldiers can also utilize light to illuminate the battlefield at night and expose enemy movement under the cover of darkness. This can be done with illumination rounds fired from a mortar cannon, handheld grenade launcher, or even flares that light up the battlefield from the sky.

As you progress in your journey, you too will search the darkness and light up the battlefield with God's illumination.

Our job as spiritual warriors is to expose the secrets and the darkness--and bring them to light.

> Luke 8:17 NLT says, "For all that is secret will eventually be brought into the open, and everything that is concealed will be brought to light and made known to all."

This is what a light bearer does. He illuminates the battlefield.

Light As a Weapon

Did you know light can be an offensive weapon? During a tactical flashlight instructor course I took, they discussed how modern high-lumen LED flashlights could be used to disorient, blind, and distract an attacker. In the same way, your light can be an offensive weapon.

The Apostle Paul gives us an example of the blinding power of God's light. Before his conversion to Christ, Paul was on the road to Damascus when Jesus appeared to him.

> Acts 9:3-4 of the Message Translation says, "...He was suddenly **dazed by a blinding flash of light**. As he fell to the ground, he heard a voice: "Saul, Saul, why are you out to get me?" (emphasis added)

Jesus tells us in John 8:12 that he is the "light of the world" There is light in the truth of God and darkness cannot exist in the light.

> 1 John 1:5 KJV says, "This then is the message which we have heard of him, and declare unto you, that God is light, and in him is no darkness at all."

So, while the enemy remains in the shadows, you warrior, will walk in the light.

Now as a police officer who experienced darkness, death, tragedy, and PTSD, I know what darkness looks like. I've seen it and tasted it, as many of you reading this might have yourselves.

But take heed, because ...

> Isaiah 9:2 NLT says, "The people who walk in darkness will see a great light. For those who live in a land of deep darkness, a light will shine."

This verse is a promise that if you are walking in "deep" darkness, you will see the light. Perhaps this is where the expression comes from, "there is a light at the end of the tunnel."

A tunnel is a dark place and sometimes it can go on for what seems to be a long time. But God's word guides our steps through the tunnel. God's word is the light at our feet.

> "Your word is a lamp to guide my feet and a light for my path." Psalm 119:105 NLT.

But we talked before about that spiritual light a believer emanates. It is visible to other believers and because of that people will begin to be drawn to that light.

Now, I know I talked about noise and light discipline. But we are speaking about spiritual light. All believers have been chosen to be bearers of God's light. You aren't given the light by God to conceal it even if it means exposing yourself to the enemy.

> Jesus says in Luke 8:6 NLT, "No one lights a lamp and then covers it with a bowl or hides it under a bed. A lamp is placed on a stand, where its light can be seen by all who enter the house."

But remember, when you bear that light, you may draw unwanted attention or persecution from the world. The world may start to see you as an enemy, and it will hate you. But Jesus said, ...

> "If the world hates you, remember that it hated me first." John 15:18 NLT

Don't let that stop you, though, and force you to cover your light. Remember there are times to use the light offensively. We cannot let the world try to stomp out our light.

The current battlefield in culture, schools, and social media doesn't like the light which is the truth of God. It finds the light offensive.

A good example that I can give is a person who is sleeping in a dark room when someone suddenly flips on the light. It makes people angry, doesn't it? Not only that but it's startling and it takes time for your eyes to just.

> 1 Corinthians 2:14 ESV says: "The natural person does not accept the things of the Spirit of God..." Nor can they understand them.

We must remember when witnessing to the world that the truth of Jesus Christ may at first seem startling or offensive to people because they are used to living in the darkness.

It would be nice to have a spiritual dimmer switch, so to speak, to gradually expose them to the light. The best way to do this is to take the advice of ...

> Colossians 4:6 ESV, "Let your speech always be gracious, seasoned with salt, so that you may know how you ought to answer each person."

Another way of saying this is to make your light palatable and easy to digest, or as the NLT version says *"attractive."*

Sometimes it's not the best tactic to throw a spiritual flashbang at people you are witnessing to. Yelling "SINNER REPENT!" and being overly condemning can cause others to cover their eyes and run from the light. Instead, calm your evangelistic SWAT team, and take a softer approach when it is needed.

Godly Grief

We must remember as warriors some people sin by acceptance and some because of ignorance. But all of it is because of the flesh.

> Proverbs 4:19 KJV says, "The way of the wicked is as darkness: they know not at what they stumble."

There's the main difference right there between people of the world and Christians. When someone finds Jesus, the internal light switches on and their spiritual night vision now allows them to see in the dark. Previously they did not know what they stumbled over.

That doesn't mean saved people don't sin, but now they are convicted and aware of it. They realize what wrong looks like and they now have the God-given desire to fight against their flesh. Conviction is a good thing. It's as 2 Corinthians 7:9-10 ESV describes as "godly grief."

> "For godly grief produces a repentance that leads to salvation without regret…"

Conviction is what makes you feel sorry, it's like a spiritual conscience, and although it produces "grief", it leads you to repentance.

You see, someone who sins continuously and has no conviction about it is walking further and further into darkness. Before you know it, they can't even see *what they stumble over* anymore. But some people choose to live in the darkness. And even when the light comes, they prefer the darkness (John 3:19-20)

If you are in darkness, the light will come but you have to go to it. And doing so may be uncomfortable at first--just like a light switch being suddenly switched on in the darkness. The light will come, but likely first with conviction.

Your spiritual eyes have gotten used to the dark. Your pupils have dilated and the light of truth and the light of conviction will sting when it first hits.

When we are convicted and confronted by our sin, this can be an emotional moment. We hurt because we hurt God, and now instead of going along with our flesh, we war against it. That's right, we war against our own flesh when we are saved.

> Galatians 5:17 NIV says, "For the flesh desires what is contrary to the Spirit, and the Spirit what is contrary to the flesh. They are in **conflict** with each other, so that you are not to do whatever you want." (emphasis added)

If you aren't warring with your flesh, then you are likely going along holding hands with it.

I will never promise you that living in this body, on this earth won't be a fight. Believe me, I also would like to be free of sin, the sting of conviction, and the pain of letting God down.

I imagine the closer one is to God *and* the more one understands God's heart and the scripture, the easier this becomes and I pray that for all of us...myself included.

I think the mistake we fall into as a clever snare of the enemy is believing that there are people out there that don't war with their flesh because we may perceive them as "holy." At times, I have looked at pastors, televangelists, prophets, and preachers with the thought, *"I just want to be at their level of spirituality and righteousness."*

The thought *"they must be so holy and favored by God"* assumes they don't also struggle with matters of the flesh as the rest of us do.

I think this a trap of false expectation of what spiritual warfare is...that is, to assume there is a level of holiness where you no longer war at all.

I believe to some degree we have to remain on our spiritual guard as a reminder that the enemy is always lurking until the day Jesus returns and we go up to meet him in the clouds. It's the age-old principle in both the military and law enforcement environment that complacency kills. And so does spiritual complacency.

I remember being surprised at a young pastor of my church saying that he had a porn accountability program and blocker on his devices during one of his sermons. He told the church *"because I know my weakness."*

I remember thinking, *"wow, did he just admit he struggles with lust to the entire congregation?"* But you know what? I struggle with lust...or rather now I battle or war against it, to put it on better terms. I use different kinds of tools and guidance to fight against it.

You might say, why am I listening to this guy then? He is weak... and you know what you're right.

> "For when I am weak, I am strong."
> (2 Corinthians 12:10).

We are strong in Jesus in our weakness, because we are flesh and fallen--but that never excuses our sin either. That doesn't mean giving in, it simply means going to war against it.

Remember only Jesus was without sin. But here is the final point in this chapter. Jesus knows the battle and the struggle we face. Hebrews 4:5 NLT says:

"This High Priest [Jesus] of ours understands **our** weaknesses, for he faced all of the same testings we do, yet he did not sin."(emphasis added)

That doesn't mean you're expected never to mess up, it just means Jesus knows temptation and He understands **our** weakness. Jesus was still the Son of God regardless of becoming flesh and dwelling among us. (John 1:14)

So understand that, as you continue to walk in a dark battlefield, be someone who strives to carry the light.

> " Instead of going along with our flesh, we war against it."

4.
Know Thy Enemy

This expression might sound like a bible verse, but it's not. It's from Sun Tzu, The Art of War. Sun Tzu was an ancient Chinese military general and philosopher. He said, "Know your enemy and know yourself and you need not fear the result of a hundred battles."

This is sound advice for any tactician and especially a Christian. You cannot defeat an enemy if you don't know who the enemy is and how they operate.

I was once speaking to someone about the devil and I referenced him as "the enemy." The person I was speaking to responded back to me rather confused and said, "what enemy?" How can you fight something you don't know or believe exists? Knowing the identity and the strategy of the enemy is important in any conflict.

The principle even carries over to business and sports. A company knows its competitors, its marketing strategies, and its customers. A boxer will watch video footage of his potential opponent and sports teams analyze the plays of the opposing team.

On the military and law enforcement side of things, they conduct reconnaissance and surveillance on enemy forces and criminal activity.

The first thing you need to realize is who you are fighting. Who is the enemy? Well, we briefly talked about this in a previous chapter. The enemy is the father of lies, the fallen angel, Lucifer, who controls the principalities of darkness in high places. (Ephesians 6:12)

> "This great dragon—the ancient serpent called the devil, or Satan, the one deceiving the whole world—was thrown down to the earth with all his angels." Revelation 12:9 NLT

Now if this were a military operation, we would plan our mission around what intelligence we had on the enemy. In this case, our intelligence comes from the Bible but also in ways evil has manifested itself throughout history.

During and after the Cold War, for example, the Soviet Union was the main focus of the US Military. There were entire field manuals on Russian military equipment and weapons.

Like myself, my father before me had been an Army Officer. He served primarily during the 1980s. While cleaning the garage one day, he pulled out and found stacks of Army field manuals about the soviet military, one of them being FM 100-2-1 "The Soviet Army Operations and Tactics."

Even when I was in the military decades later, we were still shooting at the same little green men pop-up targets that had earned the nickname "Ivans" from my father's days. I guess old habits die hard. My point is to look at how the Army prepares for war with an adversary by dedicating its energy against them.

On the smaller scale level of platoon and squad level operations, a common acronym used by Army leaders is METT-TC.
- **M:** Mission
- **E:** Enemy
- **T:** Terrain
-
- **T:** Troops
- **T:** Time
- **C:** Civilian considerations

For our purposes, we are going to take what applies to our battle.

As Christian warriors, we have a mission to spread the gospel of Jesus Christ, to go and baptize people in His name and make disciples of the nations (Matthew 28:19).

> Jesus says in Matthew 10:8 NLT, "Heal the sick, raise the dead, cure those with leprosy **[disease]** and cast out demons. Give as freely as you have received!"(emphasis added)

Most believers are unaware that we have this mission or are even capable of these things. But Jesus tells us in John 14:12 NLT,

> "I tell you the truth, anyone who believes in me will do the same works I have done, and even greater works..."

So we know the overall mission of the church, but this isn't a peacetime operation, folks--this is war. There is an enemy and he is looking for people to destroy (1 Peter 5:8).

We will talk about 1 Peter 5:8 more in detail later on but first, let's talk more about the enemy forces.

The Enemy's Mission

In combat, we want to look at the size and type of enemy we will encounter. We know as Christians we will deal with demons directly but we also have to be aware of people under the influence or being used and tempted by them as well. It could be a boss, a co-worker, a stranger...or even yourself. Have you ever been in a scenario where you feel like someone was unfairly or unjustly attacking or persecuting you?

Maybe it was a coworker, schoolmate, or sometimes it is even a "friend." The enemy can use people to influence or attack you.

What? Ok listen, I am no expert on the subject, nor do I wish to be, but people can be both possessed or influenced by evil forces. If you are Christian, and you accept Jesus, I don't believe it is possible to be possessed but you can still be tormented and influenced.

What is the enemy's mission? To take you away from God, to destroy your marriage, your family, and your life. The enemy hates mankind, he hates you. But, why?

His back story doesn't really matter, but as a created being himself, he was an angel of God, apparently very beautiful but he wanted to exalt himself to the level of the Father. When God created man we stole his glory (Ezekial 28:17, Isaiah 14:3). Now we see why things such as pride and jealousy are based on evil because those things are part of his nature.

Here is a quick piece of advice. There is no praying for evil forces, and there is no negotiating with them. Got it? Let's move on.

What are the enemy's weapons? Well, our enemy's main weapons are deception and lies. The enemy we face has multiple ways or "avenues of approach" in military terms, to come at us. He often uses terrain and timing to strategically attack our areas of weakness.

How the Lion Hunts

Let's go back to 1 Peter 5:8 and talk about how the lion hunts prey as an example. The scripture says,

> "Stay alert! Watch out for your great enemy, the devil. He prowls around like a roaring lion, looking for someone to devour."

Why would the Bible point to the lion's hunt? Modern society very rarely encounters lions unless they are in a zoo, but ancient shepherds did. There is a famous encounter in the Old Testament where young David kills a lion attacking his sheep. The Bible points to David as a warrior and the enemy as a lion in this incident. The enemy, the adversary, is cunning, and his tactics mimic that of a lion's hunt.

Lions can hunt in multiple ways, and they adapt their tactics based on size, strength, and how aggressive their prey is.[2] Lions will attack smaller and weaker prey directly with little to no strategy because there is no real danger to the lion.

People of the world that don't have Jesus are no challenge, they embrace sin, and because of this sometimes, they are even left alone. There is no sport in it for the enemy. Lions however *prefer* bigger prey…as does the enemy.

The enemy doesn't want the small prey, he wants the ones that are most dangerous to him, the evangelists, the prophets, and those seeking Jesus with the most zeal. The ones destined to fulfill the missions and ministries of Christ.

Oftentimes, these very callings of God are born out of struggles, attacks, and from fending off lions.

The enemy, like a lion, doesn't always attack larger prey straight on, it must lie and wait, or use other creative tactics, hunting in packs and using other lions to drive the prey to where it is hiding. In the same way, the enemy (Satan) will drive you into a specific situation with temptation. He will present you with people, places, images, or circumstances to lure you into the brush.

The body of Christ should travel together as a herd. Together, like elephants in a group, we pose a threat to the lions who try to attack. So what does the lion do? The Lions try to separate individuals from the herd, from the body of Christ, to get them alone (or to feel alone). But soldiers never fight alone, and neither should you, Christian warrior.

---

[2] "How Do Lions Hunt? Know Their Tactics" Snow Africa Adventures, https://snowafricaadventure.com/blog/hunting-strategy-of-the-lions/

Troops and Terrain

This is where the "Troops" is METT-TC comes in. As a warrior-leader, you need to know what resources and other fighters you have available at your disposal. I'd like to take this time to say that if you aren't associated with a local Bible-based church, you should find one and continue your growth in fellowship with others.

Something I have found interesting is that in recent culture, teenagers are now referring to their group of friends as a "squad". There are even t-shirts and memes using the phrase "squad goals."

A squad in the military is a group of six to ten fighters. Warriors don't fight alone, even at the smallest level. They are always at the minimum in a team. Being part of a small group at church, for example, would be akin to a squad whereas the whole church congregation is a company or battalion.

Together communities of Christians make up large divisions and the body of Christ as a whole is the entire Army...of God.

Now talking about enemy strategy can be intimidating, trust me, but you have to view it as if God is in on the planning process. God is like your commander, and you are giving him a briefing.

God wants to see what intelligence you've collected and how you will use it. Battlefield intelligence is vital and just as important as enemy intelligence. You need to be aware of both because the battlefield can be manipulated by the enemy.

The enemy uses barbed wire, and mines and sets up certain areas for choke points, roadblocks, and kill zones. These are meant to draw you into an ambush. The terrain is often utilized for the advantage. Hills and buildings are used for high ground. Natural or man-made funneling of valleys, woods, and narrow streets are all examples of terrain ambush points.

Have you ever heard the expression "you're in the thick of it"? The thick of what? The woods. How about the "valley of the shadow of death"? (Psalm 23:4) Or the expression you're headed down a "one-way street". They all reference terrain that can symbolize traps and danger.

I want you to see where I am going with this analogy because we are going to discuss these things as we move further along and it will make more sense when we apply it to the spiritual fight.

Time

Now let's talk about time! I probably could have devoted an entire chapter to time--but, alas, the hour is near and the *time* is short. Soldiers need to know how much time is available for them to perform an operation. They need to know the constraints and how to be efficient and effective within a given window of opportunity.

Friends, Jesus already has told us that He is returning soon and we should be prepared for that time. I have a feeling based on the way things are going in the world and the urgency God has given me in this book that "soon" is likely much sooner than we think. I don't know if that's a grammatically correct statement, but I am going with it anyway.

In the meantime, we have to get to work because no one knows when Jesus's return will be. (Mark 13:32). But scripture says that when it does come, it will come upon us unexpectedly like a *"thief in the night."* (1 Thessalonians 5:2).

Therefore we have to be careful and wise with our time. Or as scripture says: *"Making the best use of our time."* (Colossians 4:5, Ephesians 5:15-16)

Psalm 90:12 speaks about learning to number our days. I know this sounds morbid, but don't look at it as the number of days until you die. You aren't scratching days down on a prison wall until execution. You are simply taking inventory of your time in order to accomplish your mission. You are placing value on your life.

And regarding the end times, and the return of Jesus, the angel tells the apostle John in Revelation 1:1 that things will "soon take place." So you need to allocate your time appropriately and make sure you are ready during these unprecedented times.

> Jesus warns in Luke 12:35 ESV, "Stay dressed for action and keep your lamps burning."

I like the ESV version because it uses the word, "*action.*" Just as a soldier is ready for action or battle. Jesus also tells us parables using both the bridesmaids and servants in the gospels when illustrating the same point. That is to warn us to be ready and prepared for when the bride or the master suddenly returns home.

Knowing your enemy is necessary for victory. The Army loves its acronyms and METT-TC is one of the most used analyses in the military decision-making process (MDMP). Yes, another acronym.

Everything in life is a decision. When to get up, what to eat, how to dress etc. These decisions start off simple but end up with moral consequences. For example, that might be deciding if your outfit is appropriate (cough...women).

Later on, you might be at work and be tempted to gossip, swear or lust. These are all decisions and they must be weighed using a process.

I believe God wants you and I start looking at this process tactically. At some point throughout the day, the enemy will place temptations and obstacles in your path, but if you start to view things under the MDMP process, you can plan for these things ahead of time.

It's all about understanding how the enemy operates and what you can do to fight it.

5.
Principles of Security

The concept of "Pulling Security" is a basic but important tactic in the military. In essence, it means, stand guard, watch, and cover. Let's give an illustration.

A squad of soldiers is patrolling in the woods. The squad comes to halt and the squad leader takes out the map. The squad is stationary and now vulnerable to attack.

Soldiers know when they halt they have to automatically "pull security" which means protect the squad from attack. This is usually in the form of a 360-degree protective circle or "security". Soldiers in the squad will take a knee or go prone (lay down) facing outwards creating protection around the squad from attack.

What's the principle here? We patrol together, we protect each other. We watch each other's backs.

When we patrol the worldly battlefield, we move as one body, one unit, alert with heads on a swivel, and pulling security. What does that mean as a spiritual warrior? It means watching for attacks and praying for one another. Let's look at what the Bible says about 360 degrees of prayer security:

> "Pray in the Spirit at all times and on every occasion. **Stay alert** and be persistent in your **prayers for all believers** everywhere." Ephesians 6:18 NLT (Emphasis added)

> "...pray for all people. Ask God to help them; **intercede** on their behalf, and give thanks for them." 1 Timothy 2:1 NLT(emphasis)
>
> "Confess your sins to each other and pray for each other so that you may be healed. The earnest prayer of a righteous person has great power and produces wonderful results." James 5:16 NLT

This is prayer security. But why confess my sins to someone else? First, so they know how to intercede for you in prayer, and second as an exercise in humility.

God wants us to share our burdens and our sins with others so we can pray together for one another. Again, it's fellowship and fighting together.

This works even better in a group where people share similar sins such as in faith-based addiction and recovery groups.

But there are some basic tactics you need to learn first before you can move as a squad. You need to learn individual movement techniques or as the Army calls it, IMT. IMT is learning how to move as an individual soldier on the battlefield.

Before you can move as a member of a team or squad, you must first understand basic soldiering skills.

These are things such as low and high crawling and principles of cover vs concealment. What is crawling? It's moving along on your stomach. How do I describe this? If you were on the battlefield and bullets were flying overhead, you'd want to be as low as possible...moving towards something to protect you... that's crawling. It means keeping your head down.

Cover vs Concealment

Warriors need to know simple tactics such as seeking cover, moving from different positions of cover, staying low, and avoiding exposure to incoming fire. You never want to be out in the open exposed to the enemy.

Let's talk about exposure for a second. If you are out in the open, you subject yourself to spiritual sniper fire. What is spiritual sniper fire? It is purposely exposing yourself to precision attacks directly from the enemy. A sniper locks onto his enemy putting him right into the crosshairs.

When you expose yourself to things of the darkness in television, music, games, or books, you are walking into the enemy's crosshairs. A sniper is able to deliver pinpoint and accurate hits when you expose yourself this way. Always move from cover to cover.

As a Christian, this means moving through the battlefield of this world, under the cover of God. We are constantly navigating the battlefield through this life so we must know when to keep our heads down from flying arrows and when we need to rush to cover.

> Psalm 91:2 NLT says, "This I declare about the Lord: He alone is my refuge, my place of safety; he is my God, and I trust him."
>
> Verse 4 says: "He will cover you with his feathers. He will shelter you with his wings."

So, we are looking at God providing us cover, protection, and shelter. There is a very basic concept in the military known as "cover vs concealment". It's important to recognize that cover is greater than concealment. Cover can conceal you, but more importantly, it can protect you. A bush, for example, might conceal you but provides no protection. A wide tree or a large boulder, however, will do both.

God needs to be our place to cover. He is our "rock" in which we take refuge (Psalm 18:2). Psalm 91:2 says "He alone" is our cover. Not all cover is equal, and cover can disintegrate when hit with incoming fire. Some cover is temporary but better than no cover, but God is our rock and ultimate cover.

God is like standing behind layers of impenetrable steel. When out in the open taking fire, you want to seek the best cover you can. If you're in an urban environment, for example, a vehicle provides more cover than a Post Office mailbox. Ideally, though you want something large, solid, and immovable--like God.

The problem is that we put our trust in weak cover most of the time. When we take fire our instinct is to run to the nearest cover, but that might not always be the best option, because not all cover lasts.

What do I mean? Let's take health for example. In our world, we place trust in medicine, diagnosis, and men (or women). Now I want to be clear, I am not saying to dismiss medicine and science, I am telling you not to rely on them as your sole cover. Let me get personal here for a second.

I felt as though I wasn't qualified to write a book on spiritual warfare because I too struggle with spiritual and mental health issues. I have told you I have struggled with PTSD, anxiety, depression, and OCD intrusive thoughts. I also have been on and off medicine for these issues.

I believe there is nothing wrong with medicine, as it can ease symptoms but it is not to be your sole form of cover. Are following me? The same goes for treatments and doctors. These things help, but they aren't the be-all and end-all.

Mental And Spiritual Attacks

Now, as someone who struggles with mental health issues, the question becomes whether this is a spiritual battle, physical, or both. It depends on which side you take. I believe as it concerns mental health, we have to be careful not to be overly on one side or the other.

Some pastors and preachers might say "it's a demon" and it's all spiritual, and "you're a heathen and that's why you suffer." On the other hand, the world and physicians will tell you it's a biological, chemical, or genetic problem.

Remember what I said earlier, the mind is flesh. The mind is subject to sickness and decay just like any other part of your body. This is evident with age-based mental illnesses such as Alzheimer's.

Most anxiety issues, though rooted in spiritual issues, can cause physiological symptoms. Would you take medication to alleviate symptoms if you had a stomach problem?

I believe it was the late prophet, Kim Clement, who said regarding people who committed suicide that their mind was ill just like any other part of the body that gets sick.

Now suicide is never an option because you are robbing God of his purpose and will for you. You are robbing others of your presence and impact and you are robbing yourself of spiritual growth and possibly even salvation, depending on the state you are in at the time.

I don't want to get too into this topic, I just want you to understand, God has a divine purpose for your life. There are things you need to fulfill and people you need to meet. God put you here for HIS determined length of time to accomplish HIS goals, even if they seem unimportant or mundane to you. We always have to put God first even above our own discomfort and suffering just as Jesus did.

It reminds me of Jesus's plea in the Garden of Gethsemane, ...

> "Father, if you are willing, please take this cup of suffering away from me. Yet I want your will to be done, not mine." Luke 22:42 NLT.

Your troubles could end up being someone else's rescue story. You just don't know what God is doing. Think about that for a minute. When we suffer it becomes all about us no doubt, rarely do we ever consider the pain has a greater purpose.

After going through my PTSD trials, I began to get better, and eventually, I was off all of my medication. I went back into the Law Enforcement world as an inmate transport officer working at a county jail part-time to see how things went.

After a year into the job, I was attacked again, this time, it was different. The thoughts were non-stop and frightening. In fact, the day they came on me, I had been transporting two murderers to trial in my transport vehicle. I don't believe the timing and proximity of the attack were unrelated. Long story short, I had to go back on medication. I had already published my other book *Invisible Wounds* and was promoting my successful recovery, thanks to God.

I felt like I failed, as a believer, a Christian, and a role model. I was so embarrassed that I was going to pull my book down from all areas of sale domestically and internationally.

Three years prior to that my faith went through a massive trial and subsequent growth. I had been writing prophecies and devotionals. How could this have happened? I knew the scriptures now, I knew God was involved in what transpired, but how could this be God now, in this? *"God, why are you allowing me to be attacked?"* I asked.

I felt like I had no one to turn to this time because all my options were already exhausted. I had been through counseling, and I had grown faith and knew the word so how come I was having trouble fending off the attacks?

I am starting to believe that these periods that we go through are a test of strength in people's lives. I am also starting to believe that because the Lord called me to write this guide I truly had to experience the darkest forms of spiritual attack.

Anyone who has suffered from OCD and intrusive thoughts knows they can be very scary, dark, and perverted. Imagine living in constant fear of yourself. Intrusive compulsive thoughts can attack anything from your sexuality and sanity to your religious beliefs. Intrusive thoughts tend to be much worse for a believer simply because of your high moral compass. These thoughts key in on the exact opposite of your core beliefs.

I don't claim to be a psychiatrist, but I believe the way these thoughts work is they focus on your deepest fears. These thoughts are not things you wish to do, they are what you are most afraid of doing if that makes any sense.

The strange thing is that when these attacks come they actually feel foreign as if something is literally attacking you. Because most of the time, if those thoughts come under a healthy thought process, they mean nothing or seem ridiculous outside of moments of a severe attack.

I don't know how or why this occurs but there is definitely a spiritual element to it. I can one hundred percent attest to that. And perhaps that is what God is trying to show me and you through this all this suffering.

But this brings us back to what I was saying before about mental health as both mental and spiritual. When I came under attack the second time, I had gone to church one Sunday with my kids. I was seeking spiritual comfort and help. I began breaking down and crying when I was approached by a church greeter. He sent over the associate preacher.

When I opened up to him about what was going on he said, "you know, not every problem is spiritual. Maybe you should see a psychiatrist." I don't believe this is an appropriate response from a pastor. His job is to address our spiritual needs. So we need to be careful how we handle those who are under spiritual attack and not discount it as merely medical.

During this time the enemy was telling me many lies about myself and my faith. Lies will try to keep you in the dark and from moving forward. As I mentioned in the last chapter with the lion's hunt, as you become stronger in your faith, the attacks become more sophisticated.

Now it's important to note that spiritual attacks aren't just dealing with mental health. Lust is also a large area of attack for some. And it should be no surprise that I too have struggled through my life with lust. This seems to come with the territory, and it is no surprise that spiritually attacked individuals often battle in this area too.

Most people would be surprised to know that a lot of Christian men struggle in this area. Realistically a lot of men, in general, have a problem with lust. The difference is that non-Christians don't *struggle* with it, they embrace it.

I purposely choose my words carefully there to illustrate the point of the difference between being convicted and battling with something versus just having a "problem" with it. The world doesn't necessarily see a problem with certain things anyway.

Now again, there will be people who say I am disqualified from talking about spiritual warfare because of these vulnerabilities. My answer is so which sin do you struggle with? Because we all struggle with problems and sins of the flesh. Remember Jesus said in John 8:7 that the one free of sin can throw the first stone at another.

Any teaching that says you can be one hundred percent free of all sin is false. This is not possible in this body or in this world. Again, this is not a free pass to sin. Instead, be convicted of your sin... and at war with your flesh.

So, I guess the question is how do we protect ourselves? Well, as I just talked about, we make God our primary means of cover. But there are some other tactical things we can do as well related to maintaining security in our lives.

Establishing A Perimeter

We should establish a perimeter. A Perimeter in the military and law enforcement world is a way of maintaining control of what comes in and out of a specific area. It is used to contain and protect. In the same way, we need to set a perimeter around our spiritual lives. We do this by setting boundaries of what comes in and out.

There are going to be certain people, places, and things we need to keep out of our perimeter.

Let's start with friends. You may have heard the expression "your circle of friends." Who are you letting into your circle? People in your perimeter have access to and influence your mind and soul. Are your friends with God?

I am not saying you can't hang out with non-Christians, but you need to consider if certain relationships are beneficial or harmful to your faith. If you are recovering from alcohol addiction for example, should you be hanging out with your friends who drink?

What about friends who swear, talk dirty, or gossip? What impact will they have if you let them in your circle? Scriptures tell us not to use foul language, slander others, or gossip (Ephesians 4:29, James 4:11 2 Corinthians 12:20).

Your perimeter should be like a filter or a net. It needs to be selective. Now if this were a military operation, your perimeter would have early warning devices such as trip flares or alarms. These are devices that notify us when the enemy is probing around our perimeter. For our purposes, we can call these spiritual red flags.

You should start to notice when things are attempting to infiltrate your perimeter. The Holy Spirit will start to set off alarms and flares of conviction. Maybe you are scrolling social media or watching a television show and you start to cross into sexually explicit material or something that is going to trigger or tempt you. This is a perimeter alarm warning you to stop the content from entering.

The Holy Spirit is giving you boundaries and a way out of temptation. There is nothing in this world that hasn't been a temptation to others.

> 1 Corinthians 10:13 NLT says, "No temptation has overtaken you that is not common to man. God is faithful, and he will not let you be tempted beyond your ability, but with the temptation, he will also provide the way of escape, that you may be able to endure it."

This scripture plainly says most of these things you will encounter are pretty common in this world, but there is always a way out, and that doesn't mean the back door of the gentlemen's club.

There are some battles that we shouldn't enter in the first place. When you look at the METT-TC and MDMP principles, you are effectively gauging the threat to you and your forces. You wouldn't want to attack a fortified enemy bunker with only two men, for example, because the odds are stacked against you. This is just plain dumb strategically. So don't put yourself in a position where you are going to be outgunned and the enemy has the advantage.

The only way a perimeter works is if you maintain and keep it tight. If there is a weak area in the perimeter the enemy will exploit it. Instead, what you need to do is identify the weak areas in your perimeter and find ways to fortify them. This could mean adding more troops to the area or hardening structural defenses.

In the spirit, this means focusing on areas of weakness in your prayer and faith life, asking others to intercede for you, and making sure boundaries are set in place. Remember we talked about the pastor at my church with the pornography blocker on his phone and computer? This is an example of setting a boundary and perimeter but it is also an example of temporary cover. At some point, you can't just rely on only one boundary or one form of cover. The enemy could find other ways around it. Ultimately you will have to rely on God.

Set boundaries and security perimeters everywhere you go and in everything you do. Temptation and attacks will come but we want to have a tight perimeter with a decent standoff distance from keeping the attacks from reaching us and inflicting damage. That doesn't mean living in a bubble, because that is not realistic either. It means to be aware, and alert like Ephesians 6:18 says.

Remember the tactics in this guidebook are ultimately supposed to be viewed as tools for your toolbox. They are concepts and principles that when combined will help you in your spiritual fight. They work as a system not individually.

Stay alert and maintain security.

6.
Defensive Operations

I am not a sports fan. However, in football there are elements of both defense and offense to be successful. In the last chapter, I talked about perimeters and security. We are going to expand this even more as we apply it to defensive strategies when the enemy attacks. I mentioned previously that the enemy manipulates the battlefield and terrain. In war, we see examples of this on both sides of the conflict.

In defensive operations, the plan is to slow or halt the enemy's advance. In World War II, the German Army placed large metal "X" shaped objects known as "hedgehogs" on the beaches of Normandy and Omaha to slow Allied landing craft and invading soldiers. In urban environments, checkpoints and roadblocks are often set up on streets to control access and create choke points.

But just as the enemy seeks to exploit and manipulate the battlefield, so can we. The difference is our God is more powerful than any weapon of his. We have to be mindful of how we set up our defenses and not just *how* they are vulnerable but also *when* they are vulnerable.

Standing To Arms

During patrol operations, when soldiers halt for an extended period of time, they set up what's called a patrol base. It's basically a perimeter to sleep, halt, etc. It is a form of *extended* security.

That doesn't mean the soldiers let their guard down. People have to take turns pulling security as we discussed earlier, in case of an attack. This is usually done in shifts.

There are two critical points in the day when the entire patrol base is on security "standing to arms". This means everyone is up and ready for an immediate attack. The time for *"stand to"* is 30 minutes to an hour before nautical twilight--both in the morning and at night.

That's when things are under cover of darkness. This is believed to be an ideal time for an enemy attack ... when you are under darkness. It's also when most people are weary.

> Jeremiah 6:5 CEV says, "we'll attack after dark and destroy its fortresses."

Yes, the enemy Lurks and prepares for the opportune time to attack and it's usually when you are tired and weary.

> "They attack, they lurk,
> They watch my steps,
> As they have waited to take my life."
> Psalm 56:6 ESV

"Stand to" started in WWI originating in trench warfare. This is because the enemy would cross the no man's land between both sides of the war during the cover of darkness. So the order of "stand to" was given for all soldiers to fix bayonets and be alert during these times of the day in case of a charge from the enemy.

1 Peter 5:8 illustrates this in the many different translations of the Bible. The warnings used are to remain alert, watchful, and cautious of the enemy that prowls.

> The CEV Bible says: "Be on your guard and **stay awake.** Your enemy, the devil, is like a roaring lion, **sneaking around** to find someone to attack." (emphasis added)

This tactic is still enforced today and is part of patrol base operations. You see, you are the most vulnerable when you need to be the most on guard, and ready to repel an enemy attack. The enemy will kick you when you are down, and he will attack you when you are your weakest.

This can be anytime you are weak--both spiritually or physically. When people are tired, sick, lonely, bored, or even stressed out, this is often when the enemy comes around. For example, someone struggling with alcohol, pornography, or drugs might be tempted or triggered by one or more of these conditions being present.

Personally, I found attacks and temptations to be amplified myself when under these conditions. Another time which I often find myself under attack is at night and first thing in the morning, both times when I need to be at a spiritual "stand to."

Let's talk about dreams. God can speak to us in dreams but dreams are also a place where the enemy can attack. Unfortunately, most of our dreams are out of our control, but many people, including myself, have been spiritually attacked in their dreams.

Lust and fears are two big things that seem to attack during sleep because they come from both the spiritual realm and your subconscious. However, it appears we don't have much in the way of control when we are sleeping and what we will dream about.

Instead, what we can do is be mindful of what we are looking at during the day, for starters, and before bed. Now that doesn't mean you will never be attacked during sleep. In most cases when you are refraining from something like lust, you can almost expect attacks at the beginning of battling with it.

There is both a spiritual and physiological reason for this. Again, we have to understand that part of the battle is not just the enemy but our flesh as well.

The problem with being attacked in dreams means that you could be still affected upon waking. I usually find that the best way to combat this is to just get up and start your day. I also find praying and reading the Bible or a devotional helps.

I have found over time the more dedicated you are to staying pure during the day the less you will eventually be affected during your sleep.

Now in Chapter 2, I spoke about the armor of God and how the sword is the word of God. I mentioned that swords and weapons in general are both offensive and defensive. Let's focus on defense weapons first.

The scripture says:

> "No weapon that is formed against thee shall prosper;" Isaiah 54:17 KJV.
>
> "For he will order his angels to protect you wherever you go." Psalm 91:11 NLT.

The word of God and the scriptures are weapons and your faith is a bunker and your fortification on a hilltop.

Fields Of Fire

If you were a military leader in charge of defending a position or base, you would want to set fields of fire. Fields of fire are areas that you can cover with firepower with each weapon position. Ideally, you want what is called "overlapping" fields of fire. This ensures that every area of your defense is covered and it even creates kill zones of cross-concentrated fire.

I know the word "kill zone" sounds a bit harsh, but remember the spiritual enemy we are dealing with. Have no mercy on this enemy.

You want everything you have in your faith arsenal at the ready. Your defense needs to be like a spiritual machine gun with a high cyclic rate of automatic fire directed toward the enemy to cut him down in your defense.

You might be familiar with a popular cell phone game genre called "tower defense." These are strategy games where hordes of enemies which are often zombies, continuously come toward your position.

The goal of the game is to use the right combination of weapons and fields of fire to take out as many enemies as possible and stop the advance on your castle. The same concept applies spiritually.

How does this apply to you in your battle? As a Christian, obviously the word of God is the most powerful weapon you can have, but there are other things you can do to add fortification. Besides reading the word and attending church, we talked about the importance of small groups and programs.

The other thing you can do is listen to other evangelists and testimonies online. There are tons of sermons and messages online from well-respected, spirit-led individuals. Just as always be mindful because not every person professing to be a Christian is necessarily teaching the gospel. Your litmus test on these people needs to be the scriptures.

True messages and sermons are delivered from the Holy Spirit through the preacher, writer, prophet, or speaker. The Holy Spirit will begin to guide you in the right direction and you will begin to discern what is from God and what isn't.

Books are also another great resource to add to your spiritual armory. You are reading one right now but just don't stop there. Expand your horizons, and read books on all kinds of Christian topics to better your understanding of the fight we face. At any given time, I am usually reading two or three books by different Christian authors.

I personally find topics on prayer, the book of Revelation, prophecy, biblical mysteries, and near-death testimonies of heaven very enlightening. I usually stick to authors' works that I am most familiar with and writing styles I enjoy.

These topics might seem intimidating to the casual Christian but you need to step out in faith if you want to grow and strengthen your relationship with God.

Do this as you feel ready, and do not rush into topics that you think will overwhelm you. Sometimes you just need to have a better understanding of scripture first. Depending on where you are in your faith journey, there are all kinds of books and topics to suit your spiritual needs at the place you are currently in.

Calling For Fire

Now even if you are arming yourself in the word, wearing the proper equipment, and setting up tight perimeters and boundaries, there is still a chance you can be overrun at times.

This is where the calling for spiritual fire comes in. Under certain conditions in military operations, you have the ability to call in fire support. That means calling for indirect fire from artillery or mortar teams.

All commissioned officers in the Army and Marines know how to call in fire support, and there are even soldiers and special forces units that specialize in this.

Imagine you and a squad of soldiers are behind an embankment, in front of you is a stretch of a wheat field on the outskirts of a middle eastern city. There are multiple enemy fighters approaching through the wheat field. You are outgunned and overwhelmed. You need fire support. You jump on the radio and begin your request for fire:

"Kilo17 this is Zulu6, fire for effect, over."

"Grid NG187518, Over."

"Enemy fighters in the open, danger close, over."

"Danger Close" is a request given to the FDC or Fire Direction Center, that you are asking them to drop explosive ordinance VERY close to your position, aka dangerously close to where you and your squad are; because you are about to be overrun.

The prophet Elijah was an expert at calling in fire from God. He was like an Air Force Special Forces Combat Controller. These guys are the cream of the crop when it comes to calling in fire missions.

There are two instances where Elijah called upon the fire of heaven. The more popular of the two was upon Mt. Carmel, to embarrass the prophets of Baal.

> 1 Kings 18:37-38 NLT: "O LORD, answer me! Answer me so these people will know that you, O LORD, are God and that you have brought them back to yourself.
>
> Immediately the fire of the LORD flashed down from heaven and burned up the young bull, the wood, the stones, and the dust. It even licked up all the water in the trench!"

The second incident appeared in 2 Kings and was a response to Elijah being under attack. King Ahaziah had sent his Captain with fifty men to confront Elisha who had been sitting up on a hilltop.

> "Elijah answered the captain, "If I am a man of God, may fire come down from heaven and consume you and your fifty men!" Then fire fell from heaven and consumed the captain and his men." 2 Kings 1:10 NIV.

This happened two more times, killing 100 men, and two captains in total who had been sent to get him. The enemy was at the foot of the hill, and they were "danger close".

Elijah wasn't the first man of God to call in fire from heaven.

> "Then Moses stretched out his staff toward heaven, and the Lord sent thunder and hail, and fire ran down to the earth. And the Lord rained hail upon the land of Egypt." Exodus 9:23 ESV.

Sometimes we are in a position where no matter what we do, we feel like we are about to be overrun by the enemy. We feel backed against a wall, trapped, or overwhelmed. This happens to every believer at times, but you just need to hold your position and lay your battles in front of the Lord. For He has the firepower that you need. The scripture says:

> "You will not need to fight in this battle. Stand firm, **hold your position**" 2 Chronicles 20:17 ESV (emphasis added)

> "For the Lord, your God is he who goes with you to fight for you against your enemies, to give you the victory.'" Deuteronomy 20:4 ESV

> "The Lord will fight for you, and you have only to be silent." Exodus 14:14 ESV.

When you are in danger of feeling overrun, get on your radio and call for fire. Cry out to the Lord when you can't do it yourself.

Remember we have tools that are made available to us to aid in our defense. We want to make use of all the tools that are at our disposal. Remember, Ephesians 6:11 says we want to put on *all* of the armor of God and we want to make use of all of His might.

7.
Patrolling The Battlefield

This life is like one big mission from God. If only I knew that sooner!

Up until this point in my life, I viewed the mission about me. What am I going to do for a career? Where am I going to live? What are my goals and aspirations? We can't help it, we are almost bred to be that way by society.

Every time I went for my yearly check-up at the pediatrician, he would ask me "what do you want to be when you grow up?" It's amazing how much that changed over the years and really it wasn't until college and picking a major(which I changed 3 times by the way) that I felt I had to choose.

If only I had asked the question sooner, *"Lord, what do you want me to do?"*

Now I try to ask that question every day. The scripture says,

> "We can make our plans, but the LORD determines our steps." Proverbs 16:19 NLT

The reality is that we are all soldiers in the army of God and each of us has been given a unique mission. Yes, we all have the overall mission outlined in the Gospel, but God has given everyone separate orders for individual operations and tasks we are supposed to accomplish while we are here.

You may hear people ponder philosophically "what is the purpose of life?" The answer is to find and fulfill your mission from God! Did you ever stop and think that there are certain places, conversations, and interactions you are supposed to have? These things will fall in line when you start following the will of God.

God will lead you to people for the purpose of fulfilling multiple destinies at the same time. You never know what circumstances were presented by God to bring about his divine plan for you and others.

> Psalms 37:23-24 NLT says,
> "The Lord **directs** the steps of **the godly.**
> He delights **in every detail** of their lives."
> (emphasis added)

Like the conductor of a symphony, God is orchestrating the lives of the righteous.

So, because we are warriors under the Lord's command, we need to place the mission first.

The US Army soldier's creed says, "I will always place the mission first." And we are to place God and our mission first in our lives. When you do this everything else will fall into place as well. Scripture says:

> "Commit your work to the Lord, and your plans will be established." Proverbs 16:3 ESV

> "But seek first the kingdom of God and his righteousness, and all these things will be added to you." Matthew 6:33 ESV

So our mission is to navigate and patrol this life under the command of the Lord. We know that the battlefield and terrain are hostile and enemy contact is expected. So, we have to patrol with caution.

Patrolling is getting from point A to B, moving tactically. And that is essentially what we are doing moving through an area of constant threat, like this life. When a squad patrols, they move in wedge formation, spread out, and are always ready for an attack.

Before they can even step off on the mission, there is a lot of planning and briefing that takes place. The mission is laid out in detail and contingency actions are considered if you are attacked along the way. Any tactical operation has to account for certain scenarios.

Rehearsals or dry runs are often conducted on a "sand table" which I would compare to a map drawn on the ground reminiscent of an elementary school diorama project (minus the plastic dinosaurs).

I used to have a sand table kit with laminated military symbols, and other little objects to symbolize vehicles and troop movements. After walking through the mission, the leader always conducts a "back brief" questioning the troops to make sure everyone understands their role and how the mission will go down.

If we view our mission and walk here on earth in the tactical sense we should always be planning for contingency and "what if" scenarios not *if* but *when* we are attacked by the enemy.

Navigation

We will get to common methods of attack in a moment, but first, let's talk about navigation.

If you are looking to get somewhere you have to map it out. Nowadays, we just pull up the GPS app on our phone and it plans the route for us. While military leaders have access to GPS, they still have to know how to read and plot points on a map.

Map reading and land navigation skills are essential for Army and Marine Officers. It's one of those skills that will fail you out of an officer candidate program if you can't grasp it. I've seen people in Officer Basic School get recycled and fail out because of map reading and land navigation. Unfortunately, the saying in the military "lost like a Lieutenant" has a good amount of credibility to it.

My point is you need to map out where you are going in your faith journey. You need to decide where you want to be with your relationship with God, your family, and other aspects of your life. This needs to be intentional and planned.

If you're at a mall for example and looking up at the directory, it says "you are here" with a little indicator. So you need to analyze where "here" is in your faith life and then decide where it is that you want to be.

In military land navigation, that means plotting points on a map. A map, though, is only two-dimensional. Even though military maps are topographical; looking at a bunch of lines together on a map that represents a hill is much different than seeing the hill in person.

Sometimes what we plan for looks much bigger or more difficult when we encounter it in person. Makes sense right? Ok, let's move on.

When I use a compass for navigation, I shoot an azimuth based on the route I had planned on my map. The problem with land navigation and navigating the battlefield, in general, is that it is easy to lose your way and get lost. God needs to be our waypoint and the Holy Spirit our GPS.

What happens when you go off route on a GPS? It says one of two things, "perform a U-turn " or "route recalculating." When we listen to the Holy Spirit we also have to follow its turn-by-turn directions or we will get lost. If the spirit says "turn around" or is redirecting us to a different route, then we need to listen.

> 1 John 6:13 ESV says "When the Spirit of truth comes, he will **guide you** into all the truth...". (Emphasis added)

Patrolling the Battlefield | 61

So you need to be "led by the spirit" who "bears witness" to you. (Romans 8:14, Hebrews 10:5) Anyone in the military who has done solo land navigation, knows how easy it is to miss a plotted point and get lost. It just goes to show that we can plan our direction out but still lose sight and get off course.

Ok so you have planned your route out and now it is time to start patrolling. Life is one big patrol through uncharted enemy territory. But we walk with God by our side.

Now let's get tactical again. We are patrolling in a wedge formation through the woods. You are wearing your equipment, the armor of God, and you have your weapon at the ready. The Holy Spirit is navigating you on the right route. We should be good right? Smooth sailing as we make our way to our objective.

Let's pause here for a second. What exactly is our objective?

> James 1:12 NIV says: "Blessed is the one who perseveres under trial because, having stood the test, that person will **receive the crown of life** that the Lord has promised to those who love him."(emphasis added)

The objective is standing before Jesus and hearing "well done, good and faithful servant."(Matthew 25:23) To fight " the good fight and finish the race"(2 Timothy 4:7 NIV)

But remember, you are patrolling enemy territory here on earth. The enemy will try to knock you off course with lies and attacks to keep you from finishing the race.

> "You were running a good race. Who cut in on you to keep you from obeying the truth?" Galatians 5:7 NIV.

React to Contact

So now let's get onto some common attacks. Again, remember these are military tactical analogies. At this point in the book, your goal should be automatically thinking about how this relates to your spiritual fight. You know what "cover" means, you know what "weapons" are and you know what enemy contact is in the spirit.

The first attack is reacting to contact. This is a basic infantry battle drill. You are patrolling and you come across the enemy shooting at you.

You immediately seek available cover. The next thing is calling out the distance, direction, and description of the enemy. This is identifying the enemy attack and where it is coming from as we discussed earlier. What kind of attack and what weapons are being used against you? Again, start to play this through in your head now.

The squad leader has to now decide the best course of action. Are you in a situation where you aren't equipped to engage a superior force? Should you break contact and move back? Start applying these to situations and circumstances of temptation you could face.

The other option may be to assault forward. In this case, the squad leader brings everyone on a line directing and returning overwhelming firepower back to the enemy. The goal here is suppression, to pin the enemy down.

The next move is having an assault element move around to the enemy's flank and destroy them. In this drill communication with both teams are important to avoid friendly fire.

Real quick, do you believe there is "friendly fire" in the body of Christ? Perhaps I should discuss this in a separate chapter. The Holy Spirit literally just put this on me, thank you Lord for the continuous revelations. For now, keep this concept in the back of your mind.

Let's say we suppressed the attack and defeated the enemy, now what? Normally the whole squad goes into 360 security

Patrolling the Battlefield | 63

and conducts what's called a LACE check. This stands for Liquid, Ammunition, Casualties, and Equipment

You've just been engaged in battle. It's time to check and see if your armor has been damaged or if you or anyone else has been wounded.

When we go through periods of darkness and spiritual attacks we can become wounded and our equipment damaged. Sometimes our shield of faith is cracked, our sword has been dulled, our belt is loose, and the boots of peace are worn. This too we will discuss in depth in the chapter "Battle Fatigue."

Because we fight as a team in a squad during LACE, we consolidate our ammunition, water, and equipment so everyone has enough to continue. We share what we have with and patch up each other.

This is why fellowship in the body of Christ is so important. When we fight as a team, we bleed and we suffer together. But we also lift and equip each other to continue the fight. I just find the military analogy absolutely amazing as each revelation continues to come and I have no doubt that God had me become an Army Officer and go through all the struggles I have gone through for the exact reason of writing this book.

Spiritual Hydration

This is probably a good time to remember how important it is to stay hydrated on the battlefield. Hydrated with the living water of Jesus that is!

> Jesus says in John 7:37 "Anyone who believes in me may come and drink!" From the "living water."
>
> "But those who drink the water I give will never be thirsty again. It becomes a fresh, bubbling spring within them, giving them eternal life." John 4:14 NLT

Water is a vital piece of equipment for a soldier and spiritual water is vital for survival. Water refreshes us, hydrates us, and quenches our thirst. This is because we have a thirst for Jesus and his love.

Just as the other equipment such as ammunition is shared with one another, Christians should be sharing the love and living water of Jesus as well, refreshing those who look like they are in need of spiritual hydration.

This reminds me of a story about David in 2 Samuel chapter 23. David and his troops were in the thick of battle. David remarked how he would love to have some water from the well by the gate of the city. There was a small problem though, the well was behind enemy lines in Jerusalem, occupied by the enemy. Have you ever felt like the refreshment you were seeking was under siege?

Three of David's men though went on a special operation, behind enemy lines to retrieve water for him. David's men risked their lives to make sure their leader had water. This is an extreme example, but you have to ask yourselves, would you do the same to ensure your fellow warriors are spiritually hydrated?

The military doesn't kid around with staying hydrated. Anyone who has gone through basic training in the military has probably experienced "forced hydration." This is when the cadre or Drill Sergeant has everyone stand in formation and drink an entire canteen of water. They probably don't do this anymore, though.

Listen brothers and sisters in Christ, do you know what they say about dehydration? If you are thirsty, then you are already dehydrated. This goes for your spiritual life too. The best way to stay hydrated in the living water is to drink even when you are not thirsty! And with that, another spiritual revelation was just revealed. Glory to God!

Danger Areas and Ambushes

Now after all the equipment has been consolidated, wounds are patched up, the squad regroups and continues the fight. We always continue the fight!

We took some fire, reacted, and made a decision to engage. We prevailed, we regrouped, and we overcame. When we patrol, we are going to come across different kinds of areas and terrains that could be a challenge. We have to learn how to navigate these areas because we are at risk for exposure.

Open fields and roads for example are considered "danger areas". They are "danger areas" because the squad is at a higher level of exposure as there is little to no cover and concealment crossing these areas. They are also perfect places for minefields and ambushes.

Ambushes are one of the worst attacks you come up against in the natural and the spiritual. In combat operations, walking into an ambush will cause injury and casualties, and limited reaction time. The enemy always has the upper hand when they are laying in an ambush. They have both the terrain and the element of surprise. They most likely have the firepower advantage too.

When you walk into an ambush, the enemy waits until you are directly in the "Killzone" to unleash his most powerful weapons. We are going to talk about surviving an ambush, but fear not, eventually we are going to talk about turning the tide on the enemy.

The spiritual enemy doesn't fight fair, he will attack you when you are down. He will catch you off guard.

> "They attacked me at a moment when I was in distress, but the LORD supported me." Psalm 18:18 NLT.

Military tacticians will tell you there is only one way to survive an ambush...ATTACK through it. This is a tough pill to swallow. Look at it this way, you are caught off guard by directed attacks and the only through is forward against it. That's it, there isn't much else to it.

When ambushed you have to do your best to overwhelm and suppress the enemy with return fire and forward momentum. This is where your faith will come into place big time because you are going to need it to push forward when under this kind of attack.

Similar to an ambush, you could be bombarded with indirect fire. Anyone who has struggled with intrusive thoughts of fear, lust, or anything else, knows what being bombarded feels like. What would soldiers do in this situation? They would yell "incoming", the direction of the attack, and run to the nearest predetermined rally point.

This is why you need to have predetermined rally points to run to when you are bombarded. The first of these rally points to run to is obviously God, but it could also be scripture, family, friends, a support group, a mentor, or a pastor too. The point is to have designated areas of safety when you feel like you are being attacked by incoming mortars.

Navigating and Patroling the battlefield is not easy. Even once "saved" we have to expect that there will be moments of attack, ambush, and bombardment. The closer you become to God, the more of a threat you are to the enemy.

The most important thing is to continue to view your faith walk as a mission from God. You are a warrior in his mighty Army, and you are well trained.

8.

Three-Dimensional Warfare

As we go forward, we are going to start looking at warfare from the macro level of the body of Christ. This is necessary progression as we fulfill our calling in Christ's kingdom.

But first, let's talk about the concept of the three-dimensional battlefield. The traditional battlefield is often imagined as an open field or jungle. As I mentioned in chapter 1, the battlefield can span across different types of terrain and environment.

Possibly the most complex of them is the urban environment. The military calls urban combat, MOUT, which stands for military operations in urban terrain.

I believe the urban environment truly is more representative of the spiritual battlefield we face. You see, the urban environment spans multiple levels and layers. In traditional combat, there are battle lines drawn between forces, but in urban combat, the enemy could be anywhere.

In urban operations, you have the underground level which consists of sewers and subways, the street level, rooftops, and windows, and then the sky above. The enemy could be lurking around any corner, underneath you or above you. This sounds a lot like that of a spiritual adversary.

The other thing with urban combat is that combatants aren't always clearly defined. Sometimes you just don't know who is good and who is bad.

Enemy combatants could even be indigenous people such as armed women and children. In the same way, our spiritual enemy not only lurks in the unseen world but also among us in society. Many people have given way to the enemy. They have harbored his forces and provided quarters for demonic spirits. They have become a party to demons.

Now I want to be clear when I said earlier that we don't pray or negotiate with evil. That means dark spiritual forces, not human beings who are just lost or misguided souls. Those people we pray and intercede for, even if they are our societal, political, and religious "enemies". They are still our brothers and sisters and all children of God. God wants nothing more than for them to come home to him.

Unfortunately, when people side with the darkness, they become spiritual combatants to us. I stress the word spiritual here because I don't want this to be taken the wrong way. They are merely combatants in their beliefs toward Christ and the will of the Father, not physical enemies. We still should show love to them.

Of course, our goal is to win souls from the one true enemy and gain those souls as a fighter on our side. Because of this, it is important to handle these people with firm truth, love, and grace, so we don't drive them further away from God.

Understand that people of the world are under deception and in some cases possession of the enemy. When it comes to the latter extreme, there are deliverance ministries out there that specialize in this area, but they are not for the faint of heart. This gift and calling are reserved for the spiritually seasoned and battle-hardened. Perhaps you one day will be called into this type of ministry.

Psychological Operations

It is important to understand that deception is ultimately the main battle weapon of the enemy and we are fighting a war of the mind, heart, and soul.

Any time soldiers fight in an occupied area, such as a city or urban environment, it's important for the soldiers to try to win the hearts and minds of the indigenous people.

Remember John 15:19 says we are not of this world, so you must view yourself as a soldier on foreign soil. Just as we would expect resistance from the locals, expect resistance from the world and the people who belong to it as well, for you are like a foreign soldier to them.

This goes back to the METT-TC, the C being civilian considerations in war. Unfortunately, in this battle and more so in the end times, there aren't going to be any civilians. Everyone will have to pick a side.

As I said in the very beginning, there is no sitting on the sidelines in this war. Everyone will be recruited to fight on one side or another. You will either be a part of the Army of God or the army of darkness and I will let you in on some intelligence, only one side wins and it's God's.

In the end times, things are going to get very difficult. The war will intensify and just like in urban combat, the spiritual fighting will end up going house to house and door to door. This is going to be spiritual close-quarters combat like we have never seen before.

I would argue we are closing in on this point very soon and that's why many of you, myself included, are being trained and equipped spiritually for the great battles that lie ahead.

The Lord is going to need spiritual officers to lead and command others during the times to come. Though you may be dealing and battling with your own personal conflicts, there is a greater war that will take place for which the body of Christ needs to be prepared.

As officers in the Army of God, people will look to you to lead in the confusion, deception, and fog of war. They won't know what is true or even how to discern it. But you have been through the trial and testing, and you will be equipped for the mission.

> "Praise the LORD, who is my rock. He trains my hands for war and gives my fingers skill for battle." Psalm 144:1 NLT

Three-dimensional warfare means our enemies can take many forms and come from many areas. In the very first chapter, I spoke about how on the macro level that means television, social media, school systems, and societal trends.

But really who is behind all these things, is it really just men? The answer is no and it takes us back to what I said about the principalities of darkness and how they operate over humanity in certain facets of society.

To illustrate this point further, some of the things the enemy has been attacking are the family unit, marriage, children, and gender. Part of any successful warfare operation is psychological operations against not only combatants but civilians. From the days of intercepting enemy radio traffic with fake messages to air-dropping pamphlets over the cities to locals, the purpose of "Psyops" is to wage a mental and emotional war on the populace.

Evil forces have infiltrated almost every area of society including Christian circles and the church. There are forces at work that are trying to erase everything we know about humanity, love, and our Father God. They seek to replace it with a new religion of self-idolization. Self-idolization or any idolization is inherently evil as it places God second.

But before the enemy can get you to worship him, he must get you to worship yourself and everything else but God first. It is a dangerous and slippery slope.

What we have been seeing is essentially the same attack we saw in the book of Genesis with the fall of man in the Garden of

Eden. The enemy is trying to severely disrupt our relationship and trust in God and pursue our own knowledge.

In Genesis 3:5 the enemy told Eve that if she ate of the fruit that her *"eyes would open and she would be like God."*

From the beginning, the enemy tries to get us to worship ourselves, to view ourselves "like God." This is essentially what the enemy convinced Adam and Eve.

Only when we exalt ourselves to a "god" can we then defy God by claiming that we are something other than what he created us to be.

Understand, my friends, this is not a political issue about rights; it's a Psyops campaign against humanity to defy God.

We are seeing similar tactics against women, men, and children. The enemy is trying to convince women to abandon their God-given maternal instincts to procreate and give life in pursuit of self-worship. Feminism or any "ism" for that matter is a form of idolatry that seeks to take the attention away from God.

> Psalm 127:3 says: Children are a gift from the Lord; they are a reward from him."

So why then are there so many campaigns for women not to have children or encouragement to abort them? Why is mankind pushing the concept of overpopulation when humanity and the earth are in the hands of the Creator?

Why are our children being targeted in movies, videos, and books with sexual concepts? This is done to sway hearts and minds from an early age into accepting darkness. It's a campaign to influence humanity right from birth. An indoctrination of idolization.

As warriors of God, we need to be aware of the constant and growing psychological campaign against us. The ultimate goal of the enemy is to convince Christians that sin isn't sin, the word of God is flawed, inaccurate, not divine, and outdated.

"Now the serpent was more crafty than any other beast of the field that the LORD God had made. He said to the woman, "**Did God actually say,** 'You shall not eat of any tree in the garden'?" Genesis 3:1 NLT(emphasis added)

See what the serpent did there? *"Did God really say that?"* He gets you to question the validity of the word.

> John 1:14 says that Jesus is the word become flesh who came and lived among us. Hebrews 13:8 NLT says *"Jesus Christ is the same yesterday, today, and forever."*
>
> Isaiah 40:8 NLT "The grass withers and the flowers fade, but the word of our God stands forever."

The NIV version says that it *"endures"* forever. Hallelujah, this makes my soul want to shout! The word of God endures the test of time, the fickle minds of man, and the deception of the enemy. When all else fails you can rely on the word of God, Jesus, who was there even at the beginning of creation (John 1:2).

> Jesus says in John 17:5 ESV "And now, Father, glorify me in your own presence with the glory that I had with you **before the world existed.** (emphasis added).

This should blow your mind when you really think about it, the unimaginable nature of the triune God. Do you really believe that such a God could make a mistake or allow men to alter his anointed written word? Don't buy the argument that man altered the Bible. The discovery of the Dead Sea Scrolls itself disproves this idea.

Three-Dimensional Warfare | 73

Let's talk about why men are being demasculinized. This is because the enemy knows a strong man is a man of faith, a man of the family, a warrior, and a threat.

The enemy doesn't want strong, Godly men leading the body of Christ. Instead, he wants to convince men that masculinity "is toxic." Listen, weak men are terrible warriors both in the spiritual and the natural.

It should be starting to make sense now how there is a demonic agenda against humanity. Now, do you think the church and body of Christ are immune to these things?

The ultimate goal of the enemy is to infiltrate the church and the body of Christ. I referenced scripture earlier that said many people will fall away from sound teachings in later times (2 Timothy 4:3-4).

Do you see now why spiritually you are being taught tactics of battle and discernment? Even in the physical, warfare extends beyond bullets and bandages. That is, most modern wars are fought behind the scenes in the minds of the people.

It's very hard not to compare the warfare of our world to that of the spiritual battlefield. There is a war against the body of Christ. The enemy is using men to try to shut down or handicap the church and stop the body from operating at full strength.

He will do this by demonizing the word of God and Christ in the eyes of the people.

We will begin to see what is good called evil and what is evil called good (Isaiah 5:20).

2 Thessalonians 3-4 speaks of a time before the end when there will be a great falling away and rebellion against God and one in particular who puts himself in place of God to do the bidding of the enemy.

> 2 Thessalonians 9 NLT says, "This man will come to do the work of Satan with counterfeit power and signs and miracles. He will use every kind of evil deception to fool those on their way to destruction because they refuse to love and accept the truth that would save them."

The key word is "deception." Whether, on the personal level or the grander scale, the enemy still uses lies as his number one weapon. But we are warriors and in verse 15 of the NLT, the Apostle Paul calls on us to "stand firm" and "keep a strong grip" on the teaching that we know to be true.

Now, do you see why you are here? Do you see why God has chosen you to endure the struggles you have been enduring? Do you see that learning to fight the individual battle contributes to the overall war?

A 30,000-Foot View

Your role in this battle is important. The goal here is to start seeing the battle on a greater level. Whereas before, your focus was on your battle and sector of fire as a soldier, you now need to start looking at the war as a higher-level commander. Your overall view of the battlefield is going to expand as you are called to the promotion and to do greater things in God's Army.

What the Lord is showing me is that we are in a process of building each other up. 1 Thessalonians 5:11 NLT says *"So encourage each other and build each other up, just as you are already doing."*

We are building each other up in this time so we can strengthen the body of Christ overall. The church and the body of Christ are going to need to be unified and strong going into this fight.

The enemy is going to make his way into the church and attack what we hold dear. It is going to be our job to protect the institution that God has put in place for His children and ours!

Each of us has been given an assignment to contribute to the body and the fight. You are going to be strategically placed as a military commander over a certain area in the spiritual war once you earn your commission.

When I was commissioned as Officer, it was a proud moment. As an officer, you hold an office not just a title. You are given the responsibility of leading and taking care of those underneath you. This is not a charge to be taken lightly. Looking back at it now, I was not ready as a young twenty-one-year-old, but now I am learning to entrust my calling and spiritual commission to the Lord.

> 1 Peter 4:19 ESV says "Therefore let those who suffer according to God's will entrust their souls to a faithful Creator while doing good."

And as we battle our flesh and evil for the good of God, we must learn to take that suffering and trust it to Him.

As I have said previously, we need to be aware that as we progress in our destined role, the attacks can become sophisticated in nature. We must learn to balance pride and humility and constantly learn to lean on the Holy Spirit for guidance.

Success, expectation, and failure are often areas where saved Christians can walk into a trap.

Success can lead one astray through pride, and expectation can bring about failure.

Through my own personal journey, I had placed expectations on what recovery and freedom would look like. I had this idea that I was still struggling because I had not reached a certain level of spiritual maturity or righteousness. We tend to compare ourselves to others on the pulpit or in positions of authority in the Christian world.

We think that they somehow don't also struggle with matters of the flesh. I touched on this in chapter 3. The enemy will always tell you that you are not qualified for the position or the mission. You might be thinking, how am I supposed to lead others if I can't even save myself? When Jesus hung on the cross he was mocked from among the crowd:

> "He saved others, they said, he can't even save himself." Matthew 27:42 NIV.

Although Jesus was perfect, you are not (sorry, not sorry) and the expectation of perfection is a bald-faced lie of the enemy. One meant to sow guilt and negative emotions.

The enemy has no power to disqualify you from what God has qualified you to do.

God will never tell you that you aren't good enough. Instead, He will always point towards correction and conviction.

With any promotion, even in your spiritual life, you must be humble and thankful before God.

You are nearing an increase in authority and power. Be mindful of your anointing and the calling on your life. Never let it surpass your fear of the Lord.

9.
Friendly Fire

The phrases "Friendly fire", "blue-on-blue", and "fratricide" are all names to describe the accidental killing of soldiers of the same army or allied forces.

Friendly fire is something that can happen in instances of confusion and misidentification on the battlefield or the street. I spoke about the "fog of war" previously to illustrate that the enemy can take advantage of the confusion and get soldiers to fire on their own forces. This could be accidental or the result of illusion and deception from the enemy.

Throughout the history of war, military forces and intelligence agencies have gotten good at the art of deception on the battlefield. Before the rise in cyber warfare; radio communications, morse code, and even letter carriers on horseback were used as a means to trick the opposing force into believing something false about troop movements or positions.

If the enemy is able to get you to fire upon your own forces by means of trickery then he is getting you to do the work for him.

The church and the body of Christ are at serious risk of this. As the enemy conducts heavy Psyops operations against the "civilians", Christians as well are likely to be affected. There are really a few sides to this. For one, there are false teachings making their way into the body of Christ through doctrines of demons.

The other aspect of this though, which is not as easily recognized are seeds sown of discord, disagreement, and in-fighting in the ranks. This is something that the enemy does with pride among Christians which can grow into fighting and division.

"A house divided against itself can not stand." Everyone credits this saying to Abraham Lincoln during the Civil War when actually it was Jesus that he took it from.

> Jesus said in Mark 3:24-25 ESV "If a kingdom is divided against itself, that kingdom cannot stand. And if a house is divided against itself, that house will not be able to stand."

One Body, One Church

The best way to attack any opponent is from within the ranks. We are seeing a rise in Christian in-fighting as things are heating up on the battlefield. This is really nothing new though. Man has a way from the time of the Pharisees and the Sadducees fighting over differences in religious beliefs. One of these groups of Jewish leaders believed in the resurrection of the dead, the other did not. The key word here is "religion."

Religion is a weapon of the enemy. "Wait, what did you just say, Nick?" Yes, Religion is a spirit, evil, and a mean one at that. Religion comes from man, not God. Religion is what happens when men decide to fracture and divide based on tradition and doctrine.

> Jesus said to the Pharisees in Mark 7:7-9 NLT "Their worship is a farce, for they teach man-made ideas as commands from God.' For you ignore God's law and substitute your own tradition."

Then he said, "You skillfully sidestep God's law in order to hold on to your own tradition."

Religion and division are what happens when men try to put authority and ownership over scripture and its interpretation.

Catechism of beliefs based on "tradition" and not scripture alone according to the scripture above is in itself not biblical. I will let you do some digging on your own to discover the truth on that one for I don't want to be a part of the enemy's attempt to divide only to educate what the Holy Spirit is telling me.

But let's do a brief history lesson on religion among Christians based on the truth in the bible. The Apostle Paul had much to say about this in his letters to the churches. What is the original church? The original church was the followers of "the way" of Jesus. They were tasked by Jesus to spread the news of the Gospel to both Jews and Gentiles. To heal the sick, cast out demons, raised the dead, preach and teach the "good news" of salvation and resurrection through Jesus Christ, the son of God.

Paul called, and actually pleaded with the churches for unity.

> 1 Corinthians 1:10 ESV says, "I appeal to you, brothers, by the name of our Lord Jesus Christ, that all of you agree, and that there be no divisions among you, but that you be united in the same mind and the same judgment."

> In Ephesians 4: 2-4 ESV Paul says we are to walk "with all humility and gentleness, with patience, bearing with one another in love, eager to maintain the unity of the Spirit in the bond of peace. There is one body and one Spirit—"

Paul is saying there is one body of Christ. Folks this is the biblical truth and reality against denominations. There was no catholic, protestants, Episcopals, Baptist, Adventists, Mormons, and so on.

Paul even makes the distinction in 1 Corinthians 3: 4-5 NLT that people should follow Christ first, and not focus on the teacher.

> "When one of you says, 'I am a follower of Paul,' and another says, 'I follow Apollos,'

> aren't you acting just like people of the world? After all, who is Apollos? Who is Paul? We are only God's servants through whom you believed the Good News. Each of us did the work the Lord gave us. I planted the seed in your hearts, and Apollos watered it, but it was God who made it grow. It's not important who does the planting, or who does the watering.
>
> What's important is that God makes the seed grow. The one who plants and the one who waters work together with the same purpose. And both will be rewarded for their own hard work. For we are both God's workers. And you are God's field. You are God's building."

I hate to make this chapter overly scripture heavy, but there is so much truth in the scriptures on this subject that the body of Christ would do well to see and understand it. Romans 14:19 stresses the importance of building one another up for the purpose of God.

Gifts Of the Spirit

Now in the evangelistic community, there are many people who are prophets, preachers, and healers. Thanks to media-sharing sites, we can have access to these gifted people and their ministries are being allowed through Christ to reach more people all over the world.

When I was a young believer I did not believe or understand the gifts of the Holy Spirit because these things were not taught in my denomination. I did not understand how the Holy Spirit works and manifests differently in each individual. The enemy though has allowed man to handicap the flow and gifts of the spirit among the confines of religion. The idea has been planted that those things are no longer available to believers and somehow miracles stopped manifesting after the Bible was written.

The word "prophesy" has been made akin in some Christian circles to fortune telling and the power of the blood of Jesus to superstition. But you should now know what the word says regarding these things. I am not sure what Bible these churches are reading that they don't understand or see it.

We must always go back to what the Bible says on the matter. Now, I want to be careful not to suggest legalism either, because legalism is another form of confines and bondage that doesn't allow the spirit to manifest and move in the word of God either. Legalism is another way for men to put their authority over the word of God.

Let's first address the gift of prophecy. There are many Christian teachers, preachers, and evangelists attacking the prophetic and the prophets as "false teachers." A false teacher doesn't speak or edify the scripture or Jesus, for that matter.

Revelation 19:10 says the spirit of prophecy is the testimony of Jesus. That means all prophecies must point to and have the essence of Jesus. The essence of Jesus isn't the latest in cologne and perfume, it is the spirit behind the message.

> Amos 3:7 KJV says " Surely the Lord GOD will do nothing, but he revealeth his secret unto his servants the prophets."

Prophets are servants and messengers of the Lord, who share the essence of Jesus.

Never before have we seen such growth of prophets and prophetic people appearing on the internet and at Christian conferences and ministries. Why? God's timing.

> Acts 2:17 NLT says, "In the last days,' God says, 'I will pour out my Spirit upon all people. Your sons and daughters will prophesy. Your young men will see visions, and your old men will dream dreams."

Meanwhile, some evangelists have become almost jealous of the attention and growing traction of these ministries and the gifts that are being poured out and manifested. But God has given different gifts and abilities to different parts of the body of Christ. Should we always proceed with caution when listening to a message? Absolutely! We should also always test the spirit of every message.

> 1 John 4:1 ESV says "Beloved, do not believe every spirit, but test the spirits to see whether they are from God, for many false prophets have gone out into the world."

Again consider if the prophecy has the essence, the spirit of Jesus (Revelation 19:10). Prophecy should always be positive and never negative. Negative prophecies are a sign of false prophecy and divination.

> 1 Corinthians 14:3 ESV says "…The one who prophesies speaks to people for their upbuilding and encouragement and consolation"

We must not let the enemy divide and conquer the church by using pride and jealousy to handicap the Holy Spirit because we don't have or understand certain gifts that others might have.

In 1 Corinthians 12: 4-11 ESV, Paul talks about spiritual gifts and how they are distributed:

> Now there are varieties of gifts, but the same Spirit; and there are varieties of service, but the same Lord; and there are varieties of activities, but it is the same God who empowers them all in everyone. To each is given the manifestation of the Spirit for the common good. For to one is given through the Spirit the utterance of wisdom and to another, the utterance of knowledge according to the same Spirit, to another faith by the same Spirit, to anoth-

> er gifts of healing by the one Spirit, to
> another the working of miracles, to another
> prophecy, to another the ability to distin-
> guish between spirits, to another various
> kinds of tongues, to another the interpreta-
> tion of tongues. All these are empowered by
> one and the same Spirit, who apportions to
> each one individually as he wills.

So it is important not to discard any gift that God gives you in the spiritual or any gift given to another believer.

This book, this guide, is merely a reference for all things spiritual, at the end of the day, it is going to be the Holy Spirit who you need to rely on for truth and discernment.

My point in this chapter, though. is all the Christian in-fighting I have been seeing can only be the result of the enemy fracturing the church.

There are tons of videos online from various different Christian speakers and pastors calling other preachers and pastors false teachers, prosperity teachers, and all the other new Christianese lingo that has recently come to light. It is truly repulsive actually to see how they have begun to attack one another.

Situational Awareness on The Battlefield

Online evangelists have to be careful not to fall into the trap of likes, popularity, pride, and power. This is where self-examination of pride and motives in a ministry is repeatedly necessary to make sure you, as a Christian, are not being attacked from the backdoor by the enemy unbeknownst to you.

Sometimes Christians can be under attack but not even know it. After all, 2 Corinthians 1:14 says Satan disguises himself as an angel of light.

So how do we prevent fratricide? We need to be mindful of our target and what we are shooting at. Does the person you are directing fire toward love Jesus too? If so then he is an ally, not an adversary. If you don't agree with what they are saying, point to the scripture as your argument then let the Holy Spirit deal with them. Don't make war with other parts of the body. This is something all Christians should be cognizant of.

In military operations, it is important to know where all the allied forces are positioned and what they are doing. This goes back to knowing the bird's eye view of the battlefield.

I said before that as you move up in responsibility and rank in the Army of God, you need to start looking beyond your individual responsibility as a soldier. You can't just worry about your sector and lane of fire. You need to be aware of all the battlelines and zones where enemy and allied forces are. This is situational awareness. You need to know where your troops are in relation to the enemy and one another. This is how we avoid friendly fire.

As an Army Platoon leader, I was responsible for four squads. At any given time, I needed to know the whereabouts of each separate squad, even though they technically operated independently on their own missions. This is called C2 for command and control.

Remember when I said everyone is given a different assignment and mission as Christian warriors? We talked about some of those different assignments and spiritual gifts in this chapter. What that means for you as a leader, is you need to be aware of the simultaneous operations and missions going on at the same time and their location so you don't accidentally or even intentionally cross paths.

Here's where the saying "stay in your lane" comes from. Don't cross into the mission of others because that's how crossfire happens. Instead, respect and be aware of the other parts of the operation.

Remember the Commander-in-Chief, God, has multiple troops

and different kinds of special units with different capabilities and skills deployed all over the spiritual battlefield. We will never have a full understanding of God and the big picture but we should have a basic understanding overall of the battlefield.

A military division has all kinds of units underneath it with different specialties that could be present on the battlefield. In a theater of operation, there could be infantry, tanks, artillery, military police, civil affairs, transportation, and special forces all conducting separate missions in the same area AO or Area of Operations.

Sometimes too, we are just too focused on the enemy and the target that we are all rushing at each other. This is also a recipe for friendly fire.

Identifying Allies

I am going to switch back to my law enforcement hat for a minute for a good analogy. In an active threat/shooter event, police officers are trained to pursue the enemy and take out the threat. The problem is that everyone is pursuing the same thing and sometimes they are coming from different places of entry, wearing different uniforms and there is a high likelihood of misidentification and crossfire. This is further exacerbated by the fog of war in this environment.

When I attended the Federal Law Enforcement Center(FLETC) Active Threat/Shooter Instructor Program, they spoke about the Washington D.C. Navy Yard shooting incident. Because of the area, armed law enforcement officers in different uniforms from police to off-duty federal agents had rushed into the area and the building to pursue the threat.

During the last exercise of the instructor program, they had all of us students enter a building at the same time to show how confusing and chaotic the situation could be. In the event of an active shooter scenario likely every officer with a gun would respond, including neighboring agencies and plain clothes detectives. But one can see how this could be dangerous and a recipe for friendly fire.

Let's wrap this chapter up by bringing things back to the spiritual fight. We are all pursuing God and we are all battling the enemy. We have to be careful to identify one another as fellow Christians with the same goal, same team, and same enemy in common.

In military and Law Enforcement shouting the word "Blue" while coming across friendly forces is a way we identify ourselves as allies so we are not shot at. In the military traditionally there is also a "challenge word" and a specific response password that is given, that all friendly forces know to help identification on the battlefield.

As friendly forces come in contact, the challenge word is given, and if the designated answer is not received then you know they are hostile forces.

Fellow warriors, our challenge question is, "Who died for our sin?" If the approach warriors answer the password "Jesus", they are an ally--stand down your weapons and let them be.

It doesn't matter if they are a prophet, preachers, evangelists, healing, or deliverance ministry. If they profess the truth and essence of Jesus, let them pass. It is not your responsibility to deal with them or question their mission.

10.
Warrior Leadership

Congratulations, soldier! You've been promoted! You are now an Officer in the Army of God. You made it this far and it's due time that you are considered for moving up in rank. You've been given a battlefield commission.

This happens when someone has to be promoted on the spot in combat to take over a position. Battlefield commissions are very rare, usually. Officers have to go through an extended amount of training. For me, that was Army ROTC, followed by Army Basic Officer Leader Course (BOLC).

This is a lot of training. We are talking about four years of Army ROTC or West Point in college, a 5-week intensive advance camp during CST (Cadet Summer Training), and six months to a year of active-duty combat and specialty training at BOLC depending on your branch assignment.

This was a whole lot of physical training, running, ruck marches, land navigation, and field training exercises. But all the training in the world though can't really teach you to be a leader.

I hate to admit it, but while leadership principles can be taught, leadership in itself is earned. Most newly minted 2nd Lieutenants can tell you that earning the respect of the soldiers is not easy.

But you are now a leader in the body of Christ. Consider this guide a crash course in spiritual warfare. This is like an accelerated commissioning program since you received a battlefield promotion. The Lord needs soldiers but more specifically leaders during this time to help shepherd the people. There is a call for this position, and I can feel it during this time. The Lord is raising up an army of David's for the hour at hand. I say David because he was a warrior-leader. From the days of killing lions to giants, to leading kingdoms and armies in battle, David knows a lot about spiritual and physical leadership.

David has been called " a man after God's own heart" (1 Samuel 13:14) not because of his high righteousness but because of his relentless pursuit of God. Explore the life of David and the psalms and you will see that David deals with issues from anxiety, depression, and even lust.

God doesn't expect his Officers to be perfect, He knows you have faults and struggles. He is searching your heart.

> Jeremiah 17:10 ESV says, "I the LORD search the heart and test the mind, to give every man according to his ways, according to the fruit of his deeds."

If you really consider all the people God has chosen to do great things, they were all a little messed up--so to speak. Moses, for example, had a speech problem and was a murderer. Jacob pretended to be his brother Esau and glued hair to his body. Jonah was a chicken, the disciples a motley bunch, and Paul was a real jerk.

Trial By Fire

Ok, so I am adding some levity there, but there are many more examples of people God humbled and chose to do great things. Why are you any different? No matter what your perceived shortcomings or faults are, they are not limitations that God has placed on you. God doesn't care if you are short, tall, fat skinny, or built like a gladiator. He weighs the prowess inside of you and he does this by putting you to the fire.

> Psalm 66:12 NIV says, "For you, God tested us; you refined us like silver. You brought

> us into prison and laid burdens on our backs. You let people ride over our heads; we went through fire and water, but you brought us to a place of abundance."

This brings up a hard question. Does God allow you to suffer to become what you are supposed to be? To answer that, let's look at the military. Is military training easy or hard? Does one suffer with sweat from physical and mental exhaustion when going through training to become a warrior? The answer is yes, and training is never easy. God puts us through the test when we are called to something. And tests aren't easy.

Consider it this way, Army Special Operators go through what's called SFAS "Special Forces Assessment and Selection." This is where people who desire to be the cream of the crop in the warrior world are put to the test. The Army wants to know if you can handle the rigors of special forces training and life.

Before they even begin to train you in what is known as the "Q Course" or Special Forces Qualification Course you have to pass selection. This training is no joke. Most of the world's militaries have elite forces with extremely selective and grueling requirements.

One of the most notable is the French Foreign Legion. I always wanted to be a special forces soldier or an Army Ranger at the minimum but God had other plans. I believe God has been putting me and many others through what I believe is spiritual special forces selection and assessment.

> 1 Corinthians 10:13 MSG says, "All you need to remember is that God will never let you down; he'll never let you be pushed past your limit; he'll always be there to help you come through it."

But I thought God puts us under fire and trial for refining, isn't that contradictory to other scripture?

What it means is that God will push you out of your comfort zone *but* not your limit. He knows what you are capable of and what you can endure. Again our limit isn't God's limit.

Now there is something to be said about surviving intense and grueling training or any other experience for that matter. During my military and law enforcement career, I went through several selection courses. From Army cadet advanced camp, BOLC, the police academy, firearms instructor school, and numerous other instructor-level courses. I had to meet the challenges and prove my worth. These challenges weren't always easy and a lot of times, I was concerned I would fail.

And each time I suffered mentally from anxiety, depression, and lust, I was worried I would fail. But God continues to push me out outside of my perceived limits while keeping me inside of His. That is why we put our trust in the Lord because He has got this!

> "Give all your worries and cares to God, for he cares about you." 1 Peter 5:7 NLT.

In the next chapter, we are going to analyze this more as we talk about combat fatigue. For now, though we will focus on the role of leadership as a warrior.

Leadership Attributes

Here's a question, are all leaders teachers? I would argue yes, there is an element to leadership that requires the ability to teach.

Both the ability to teach and lead effectively requires humility, accountability, and the right heart.

Accountability is a vital part of being a leader. Not all people in "management" or positions of power are good leaders or effective teachers. This is because of pride and lack of responsibility, respect for others, and self-discipline.

Consider what James 3:1 NIV says,

> "Not many of you should become teachers, my fellow believers because you know that we who teach will be judged more strictly."

This seems kinda harsh doesn't it, but what the Apostle James is speaking about is being held accountable in your position. Have you ever had a boss tell you something incorrect or wrong and then refuse to own up to it?

In the next verse, James says:

> "We all stumble in many ways. Anyone who is never at fault in what they say is perfect, able to keep their whole body in check."

James is speaking about owning your words, especially if you are in a teaching and or leading position. Not every pastor, preacher, or Christian for that matter is always correct in their ways.

Sometimes we preach or teach from a position of flesh rather than the spirit. Our personal beliefs and feelings can get in the way. This is only normal and something we have to keep tabs on. The enemy knows how to exploit our flesh and pride is one of those weaknesses that most people deal with. We tend to want people to view us a certain way, or we become very protective when questioned, corrected, or criticized. It's important to acknowledge when you are wrong about something and that points back to having humility.

Now let's talk about humility in leadership. Humility as a Christian leader means first acknowledging God for your position and accomplishments. It also means giving credit to the Holy Spirit as well for revelations and knowledge received. This very book for example is credited to the Holy Spirit.

I have always said regarding my last book *"Invisible Wounds"* that it wrote itself, or rather the spirit wrote it. Listen, I was never good at grammar or writing and always struggled during elementary and high school with writing. The last thing I ever thought I would call myself was a writer. One can only attribute this to the Lord.

King Nebuchadnezzar is someone in the Bible that had to learn the hard way what it means to be humbled by our mighty Father. In Daniel chapter 4, Nebuchadnezzar dreams of an enormous tree that is flourishing then is suddenly chopped down and destroyed. Daniel interpreted the dream for Nebuchadnezzar that because of his pride his life would follow the same fate as the tree that was once flourishing and be cut down.

One day Nebuchadnezzar was boasting about his kingdom and achievements and the Lord suddenly drives his sanity from him. Nebuchadnezzar ended up living on all fours eating grass in a field like a cow for seven years until He acknowledged the Lord.

We must consider the motive of our hearts when deciding to lead. That means leading and shepherding with an "upright heart"(Psalm 78:72 ESV) Always placing the will of God ahead of our own. We don't want to be like Nebuchadnezzar, instead, we should be much more receptive to the Lord when it comes to guidance.

> Psalm 32:9 NIV says, "Do not be like the horse or the mule, which have no understanding but must be controlled by bit and bridle or they will not come to you."

Yes, my friends, The Bible just said do not be like a stubborn mule. These are true words of wisdom for both those who lead *and* follow others.

Lastly, let's talk about responsibility. As a leader, you have a responsibility to those who follow you. Do you have a ministry of some sort? Maybe online or even through social media. I never really considered my devotional Substack a leadership position, but the reality is that my "followers" are...well...

following me. So I guess that means I am a shepherd of sorts. I should probably be mindful of the things I am writing and posting right? Now Imagine those preachers, teachers, prophets, and evangelists with large global audiences.

"With great power comes great responsibility." Who said it? Peter Parker aka Spider-Man...or Stan Lee...right? Actually, Jesus said it. Again, art imitates life...or Jesus rather.

> Luke 12:48 ESV says, "Everyone to whom much was given, of him much will be required, and from him to whom they entrusted much, they will demand the more."

Ok so they changed the words a little but they still ripped them off from Jesus. There is actually a boatload of common sayings that are basically adaptations of Bible verses. This is not necessarily a bad thing, but people would understand them more in their faith if they knew what was behind their original meaning.

Responsibility is something that as a leader can extend beyond what is even in your immediate control. Can a leader be held responsible for something he has no immediate control or direct involvement in? The short answer is yes. What goes on under your command in the military, for example, is your responsibility.

I learned this lesson the hard way as a young lieutenant while on an annual training exercise. I spoke about earlier how as a Platoon Leader, you are responsible for the whereabouts and activities of all the squads that fall under your command. While on this training exercise, the entire company was supposed to regroup at a certain destination. All of my squads were out in various areas operating independently as they should. My squad leaders knew where they had to be and when.

When the time came, the entire company was there at the designated rally point minus one of my squads. The squad leader who was a Staff Sergeant said he got his squad lost. Was this my fault or his? Well, my commander, the Captain, told me it was my fault. Did I have much control over him getting lost? No, but ultimately he and his squad fell under my command and my responsibility. Did I subsequently chew him out for getting me chewed out? Yes but in all honestly after the Captain was done chewing me out, he told me to chew the squad leader out.

It was a teaching moment for all.

Another time I was the convoy commander as the senior 1st Lieutenant for the entire company. We were driving our Humvee vehicles on a specific route for training. The trip would take a few hours and we had a pre-planned fueling point along the way. A new 2nd Lieutenant broke the convoy and decided to pull his platoon's vehicles out of the convoy to stop and refuel.

Now, this was not a combat operation, but had it been, he would have risked isolating his platoon, breaking convoy integrity, and possibly risking lives. He told me he made a "command decision" because he was afraid his platoon's vehicles would run out of fuel. We did however estimate and account for all this when we pre-planned the locations of the refueling stops. I ripped him a new one and then the Captain ripped him a new one when we arrived at our destination.

The point is there are responsibilities as leaders we need to take seriously. We can't take such a position of authority be it in ministry or even in your community, family, or church for granted. Even if you have no intention to enter into any official ministry, you likely have a family you are responsible for. This is especially true if you are a parent. Guess what? Parents have entered into a ministry whether they chose to or not.

Your children and your family are your spiritual responsible and you are their warrior leader, so lead at the front, and lead by example.

11.
Battle Fatigue

It can happen to anyone. Sustained combat operations wear on the mind, body, and soul. The constant alertness, the ups, and downs, always being on the defensive. You feel like you're constantly taking fire, being bombarded, and dodging bullets. You've lived through some trying moments and maybe even come close to death or watched others suffer and die.

The enemy's campaign seems relentless. You are tired of seeking cover and tired of just trying to survive. You want to live a normal life of peace but you have been in combat so long you're not even sure if that's possible anymore or what it looks like. I could be describing a soldier but I am describing you and I, the spiritual warrior.

We have been in a long fight and this war doesn't seem to have an end, does it? We could be like some other Christians and stare up at the sky just waiting for the return of Jesus, but the battle remains here and we have to keep fighting. We ask *"Why Lord am I fighting so hard, I am tired and weary." "I have scars, memories that haunt me, I just want to stop fighting."*

Man do I know this feeling. Even as I write this the scars of my darkest moments are there. It's like I formed scars from the scars if that makes any sense.

When I had severe PTSD symptoms, it was because I had developed cumulative PTSD from 12 years of Law Enforcement. It was a build-up of things, not one particular incident. In fact, if you do your digging, you can even find mention of me and my story in an article from *Psychology Today* that I wasn't even aware of until recently.

I don't care what people say about cops, a career in law enforcement is like one nonstop military deployment. You just go home and sleep between shifts. This is how it always felt to me. After I had left, I began to feel relief. Fast forward three years later when the symptoms returned, I had already written and published *Invisible Wounds* and was working in the local county jail as an Inmate Transport Officer. As far as I was concerned I was saved, born again, and healed.

I was off all of my medication and previously I had been on several for all kinds of issues ranging from anxiety to nightmares. I had started posting on social media how God cured me of PTSD then days later the attack came. This time, I thought I could fight it off much sooner with my faith.

But several months later, I was suffering and slipping down a deep rabbit hole of misery. I spent hours in the Bible, and in worship music, hoping it would have the same effect on me as last time, but it didn't.

I don't quite understand what happened and why. I know the enemy wanted me to believe certain reasons why and abandon my faith but I did hold on and I am still holding on.

I had been attacked with all kinds of intrusive thoughts. It was almost unbearable. I went back on medication and that particular medication is known during the initial adjustment to cause even worse fear and panic until it levels off.

The enemy would love for me to dwell on all this but Instead, I am going to arm and equip others with it. I am telling you these things but you need to know I am a credible witness and perhaps that is why God allowed these things to happen. What good is someone telling you about spiritual warfare, if they have never spent time in the trenches themselves?

Brothers and sisters in Christ, I don't know your struggles and no one should ever pretend to understand yours or minimize them. Family, friends, pastors, and counselors mean well, but they don't know what you feel.

Nobody should tell you that you are suffering because you don't pray enough or you lack the correct amount of faith. This is the lie of the enemy, which I myself believed during my second time around with attacks. I was convinced I was doing something wrong. *"You don't have enough faith,"* I thought. *"Really? I wrote a book about it, but I am not sure if it's that."* When in fact, it is my faith that continues to sustain me.

A God That Can Relate

Nobody knows what you're going through except one person... Jesus. Jesus knows exactly what you are going through to the point that He can probably even feel your pain. Do you realize that every sin and inequality was placed on him on the cross? At that point, He probably didn't even feel the pain of the railroad-sized nails in his hands and feet from all the pain of the world on him.

Jesus didn't just endure physical and mental torture at the cross but spiritual torture. Isaiah 53:5 ESV speaks of Jesus in Old Testament Prophecy several hundreds of years prior. It says:

> "But he was pierced for our transgressions;
> he was crushed for our iniquities; upon him
> was the chastisement that brought us peace,
> and with his wounds, we are healed."

I remember the first time I wept for Jesus. It was during my initial attack three years ago that led me back to the Bible in the first place. I was reading about Jesus in the Garden of Gethsemane the night before He was to die. For the first time in my life, I felt extremely sad and struck with emotion for Jesus that I began to cry. Perhaps it was because I was hurting myself and I could now relate. Or perhaps the spirit put it upon me but I know I wasn't crying or sad for myself at that moment. I was crying for Jesus.

> "And he took with him Peter and James and John, and began to be greatly **distressed** and **troubled**. And he said to them, "My soul is **very sorrowful**, even to death. Remain here and watch." And going a little farther, **he fell on the ground and prayed** that, if it were possible, the hour might pass from him. And he said, "Abba, Father, all things are possible for you. **Remove this cup from me.**"
>
> Mark 14:33-36 ESV (emphasis added), "And being in **agony** he prayed more earnestly; and his sweat became like great drops of blood falling down to the ground." Luke 22:44 ESV. (emphasis added)

I can't even begin to imagine the anticipation and anxiety. I had a full-blown panic attack before an endoscopy procedure, let alone going to be crucified with the weight of every sin and inequality crushing me.

I asked before, are Christians supposed to suffer? Or are we just supposed to suffer in order to relate to Jesus? There is plenty of scripture that speaks of sharing in our suffering with Jesus. Numerous scripture in fact in the New Testament. But everything is temporary, even pain.

> 1 Peter 5:10 ESV says, "And after you have suffered a little while, the God of all grace, who has called you to his eternal glory in Christ, will himself restore, confirm, strengthen, and establish you."

I'll take that as a promise, thank you very much.

Scripture says that those who suffer in the flesh like Jesus, cease from sin (1 Peter 4:1 ESV).

So suffering seems to be a form of purification. I would cite scripture on trials in this life but there are just so many scriptures on it and suffering that it seems more like a commandment and rite of passage than a possibility.

That's why I ask the question, is suffering necessary for growth?

Getting A Spiritual IV

How then do we mitigate battle fatigue? The simple answer is by supernatural sustainment, not of our own power. Though we might not recognize it at the time. It is always there.

> The scriptures say, "But they who wait for the Lord shall renew their strength; they shall mount up with wings like eagles; they shall run and not be weary; they shall walk and not faint." Isaiah 40:31 ESV
>
> "My grace is sufficient for you, for my power is made perfect in weakness." 2 Corinthians 12:9 ESV
>
> "I can do all things through him who strengthens me." Philippines 4:13 ESV

The above scriptures speak of not just relying on the Lord for strength, but renewal. We renew ourselves through Jesus. Hydrating ourselves with living water, renewing our mind, body, and soul.

Sometimes we just need a spiritual IV. I was attending a law enforcement Instructor training for officer-down rescue and tactical medicine. It was July and at least a hundred degrees with the heat index. We were running around, shooting, carrying ballistic shields, dragging each other, and putting tourniquets on. It was a great course.

That night I went back to my hotel room and was trying to rehydrate. Throughout the next day at training, I was alternating between sports drinks and water trying to catch up with my body, but it was too late, I was dehydrated.

Luckily, there were paramedics going through the class with me in pursuit of becoming SWAT medics. I didn't want to notify the instructor of the class, for fear of missing out but I was feeling dehydrated so two of the paramedics snuck me away on lunch break and gave me an IV. It was what I needed and what at that point sports drinks and water couldn't do. I was able to get back into the fight and finish the class.

Sometimes we just need a spiritual IV directly from God. That living water directly into our veins. There are many ways and places to get living water, but some of those are still mysterious to me.

Lately, however, I have been finding that writing like I am doing now, seems to be a form of both living water and revelation. That doesn't necessarily mean that is what will work for you too, but I believe as we continue to put God first and pursue Him, we will be led to our own individual wells as a source for renewing water.

Sharing The Load

Let's talk about what other ways there are to lessen fatigue in our walk with God through the battlefield. We know scripture says we can cast our cares upon Jesus because he already bought and paid for them. (1 Peter 5:7) We know as Christian warriors we need to watch each other's backs and consolidate ammo and equipment. That also means we carry others and let ourselves be carried when we are wounded(remember LACE).

A common military training exercise is called "buddy carry." There are a few different methods of carrying a wounded soldier on the battlefield over your shoulder, none of which are easy or comfortable. A good teamwork exercise is the litter carry, where soldiers take turns carrying wounded teammates on a stretcher for long distances at a time.

Carrying other people's burdens and wounds is arduous and uncomfortable (just ask Jesus) not to mention awkward in more than one sense. Sometimes we don't want to share in lifting others and their burdens especially when we have our own.

The team litter exercise is a good example of hoisting others over your shoulder carrying them, then later being carried yourself. It's awkward to have a stretcher rest on your shoulders and at times painful, but it makes you appreciate when others are carrying and lifting you up too.

Remember we talked about Sharing and confessing our sins to one another so we may be healed. (James 5:6) Sometimes the best way to heal is to let go and share our sins and burdens so we are not hiding them in our own darkness. That doesn't mean unloading all your negative feelings on someone. You can save that for Jesus. Instead, there are appropriate spaces for this such as faith-based programs and small groups.

Spiritual Intercession - A Direct Line

Is there a way to unload on God though? I am going to talk about something that I want you to pray about personally. This is something that can help you and has helped me with battle fatigue.

There is a scripture and belief that our soul can cry out to God when are unable to express it in words. I don't know if you have ever felt that feeling, but I have. This is the scripture I am talking about:

> Romans 8:26 ESV, "Likewise the Spirit helps us in our weakness. For we do not know what to pray for as we ought, but the Spirit himself intercedes for us with groanings too deep for words."

> "And they were all filled with the Holy Spirit and began to speak in other tongues as the Spirit gave them utterance." Acts 2:4 ESV.

Both of these scriptures reference speaking under the power of the Holy Spirit but in different ways. This is a controversial topic among Christians, one that the enemy also used to cause division. This is known as speaking in tongues.

There are two schools of thought on this, and actually, both are likely true. Many interrupt the scripture in Acts 2:4 ESV to be that the Apostles were given the ability to speak in other known *human* languages. This is true in *this* incident. However...I believe there is such thing as a personal prayer language that comes deep from our soul and that is what Romans 8:26 is referencing.

These are two separate things, but both are powered by the Holy Spirit. How do we know that all "tongues" of the spirit in the Bible aren't referencing speaking in other human languages? The following lets us know how:

> 1 Corinthians 14:2 NIV says " For anyone who speaks in a tongue **does not speak to people but to God.** Indeed, no one understands them; they utter mysteries by the Spirit." (emphasis added)

Now before you drop the book and say "oh this guy is crazy", remember tongues are mentioned as a gift of the spirit in scripture. It is a gift for you if God lets it manifest. It's to allow you to pour out our soul's deepest grievances on the Lord in a language that neither you nor the enemy can understand. It's like an encrypted battlefield transmission straight to God, the commander-in-chief.

1 Corinthians 14:4 NIV says *"Anyone who speaks in a tongue edifies themselves..."* Again, this is a personal gift.

Growing up, my grandmother, her sister, and their church friends would pray and speak in tongues, which always sounded like nonsense to me. I never truly understood this under my last severe attack a year ago.

I told you I was in deep pain; I had been previously saved three years ago and committed my life to Christ. I was doing everything to have the same experience I had reading the word during my first attack years earlier but wasn't having the same effect.

This time around, however, the pain was soo intense and beyond human words that I began to speak in tongues from a deep place of emotional hurting. I would go draw a bath to be alone with God and just let the sounds come out riddled with an undertone of sadness. It was like my soul was just crying directly to God bypassing even my own thought.

> 1 Corinthians 14:14 NIV reaffirms what I felt. "For if I pray in a tongue, my spirit prays, but my mind is unfruitful."

I encourage you to read the chapter of 1 Corinthians 14 if you doubt the validity of speaking in tongues or view it as only speaking in other languages. If you read what the Apostle Paul is writing, you can see that he is referencing an unknown spiritual language that benefits the individual but when spoken to a group must be translated for the benefit of all.

Speaking in tongues is something that I am growing in and I find it continues to grow and can provide comfort. Sometimes praying is hard and we don't know what to say, so as the scripture says the Spirit intercedes for us. There were times I had even found myself singing what sounded like a song in words and sounds I didn't understand in a familiar and flowing melody.

I don't want to pressure you into this gift, as with every gift, the Lord will give it when the time is right and as he sees fit. It is interesting though as Paul says in 1 Corinthians 14:5 NIV " *I would like every one of you to speak in tongues…*".

I would have to agree with Paul here, as I am finding it to be a tool for our benefit in battling fatigue. But as I just said, pray about this with the Lord and seek your own guidance on it from the Holy Spirit from which it comes.

Finally, the last thing we can do to mitigate and combat fatigue is to not just fight through it but instead launch a counter-offensive and that's what this book is for me at least. This book is a guide for you, but ultimately you will need to launch your own assault on the enemy and figure out what that looks like.

In the last chapter, we are going talk about just that.

12.
Offensive Operations

We can't always be on the defensive. Sometimes you just have to take the fight to the enemy and launch an all-out offensive on their stronghold. Everything we talked about up to this point was preparing you for this.

> David writes in Psalm 18:37-40 NLT, **"I pursued** my enemies and **overtook** them….I **crushed** them so they could not rise……you armed me with strength for battle……you made my enemies turn their backs in flight."

The enemy stole ground from you, he stole ground from the church, but God goes before us in battle.

> "But be assured today that the LORD your God is the one who goes across ahead of you like a devouring fire. He will destroy them; he will subdue them before you. And you will drive them out and annihilate them quickly, as the LORD has promised you." Deuteronomy 9:3 NIV
>
> Exodus 15:3 NIV says, "The Lord is a warrior; the Lord is his name."

Verse 6 says " Your right hand, Lord, was majestic in power. Your right hand, Lord, shattered the enemy."

It is time to start retaking spiritual ground and territory. It is time to reclaim your life, your family, your home, and your faith.

Advancing on the Enemy

When we are constantly under attack the best thing to do is to go on the offensive. I talked about running for cover during defensive operations and you should definitely seek shelter and protection.

But there is an expression in the firearms instructor community called "getting sucked into cover." What does that mean? It means hiding and essentially cowering behind cover until you are pinned down. The longer you remain in that position you lose the opportunity to tactically maneuver or advance out of there.

Once stuck there, it can allow the enemy to just advance on you and overtake you. At some point to have to go on the offensive. Don't get sucked into cover.

When new recruits would come to our Police Department fresh out of the Academy, I used to run them through a block of firearms instruction. One of the exercises I had them do was to simulate a car stop where they took incoming fire as they were approaching the suspect vehicle. I had them quickly return fire while withdrawing to immediate cover behind their cruiser.

Once organized together behind the cover, I had them go leave the cover together and strategically advance on the enemy. This is a military concept that most Police officers just don't understand, so it was something I wanted them to be comfortable with.

During a similar active shooter training, a group of four of us were advancing into a building. Upon entering the hallway, we started taking fire from the shooter and two officers behind us retreated into a nearby classroom.

My partner and I however advanced and overwhelmed the assailant with firepower, taking him out. We didn't even have to communicate with one another while doing this, it just happened. The common denominator was both of us served in the Army, the other two officers who dove into the room never did.

After the scenario was over, the assailant expressed how he was impressed by our approach and commented he was immediately pinned down and overwhelmed by our firepower. And why shouldn't he have been? After all, there were two of us (well technically four) and one of him.

When you're facing the enemy in the hallway, like we were, your inclination is to run, hide, and get sucked into cover. Had we all jumped into the nearest room out of the hallway, the assailant could have just run down the hallway to our room and started taking out like fish in a barrel.

Before starting this book, I got attacked again. I almost stopped and ran for cover, but I felt the Holy Spirit say "No, don't back down, push forward." So here I am 12 chapters later, pushing forward. Listen, the enemy has no issues ambushing and attacking you, why shouldn't you do the same to him?

Your mission at some point for God will involve bringing the fight to the enemy. God is using people to perform spiritual raids breaking up strongholds of the principalities of darkness. Strongholds on the personal level and the societal level as we discussed in Chapter 1. In our case, the raid involves taking back what was stolen from us.

Taking Back Spiritual Ground

So here is what it will look like tactically. We gear up with our armor of God and grab as many warriors and heavy spiritual weapons as we can. Then we devise a plan of attack using what we know about the enemy, his terrain, and his strategies. Once we are ready, we mount up and roll out.

I am going to give you a powerful illustration from my time at Army Military Police BOLC. Toward the end of the course, you set out into the field for a week-long STX or situational training exercise, that simulates a mock deployment. I was at Ft. Leonard in Missouri where they have a training city there with multiple high-rise buildings, a mosque, and a police station used for urban combat.

The culminating event at the end of the week is an assault on the city. The mission is to take back the city from the insurgents. The insurgents had taken over the Iraqi Police station and subsequently the surrounding area as well.

During the sand table and planning, we identified two high-rise buildings to set up overwatch positions for the assault on the police station. One squad would provide overwatch from the vehicles as the other squad assaults and takes a position of overwatch on building 1. Once secured, the other squad would assault building two and set up overwatch. Finally, the last two squads would be responsible for cordoning off the area and the final assault on the police station.

The mission began at night. With night vision snapped to our helmets, we rolled in on our Humvees. Each Humvee had a 50-caliber machine gun with a gunner on the roof. (This is how MPs roll!) We pull into the city and launch 40mm smoke rounds creating a wall of smoke in the air. This smoke is meant to provide a means to conceal and obscure our movements.

The enemy forces immediately fight back, but the 50 caliber machine guns on our Humvees roar with massive thunderous reports and bright yellow flashes piercing the smoke-filled air.

I am in the assault element for building one. We run through the smoke and stack on the building doorway. With aggression, we enter making our way through the dark building hallways, under green night vision eyes, clearing each room, and making our way to the upper floors.

When the building is cleared, we launch a flare into the sky to signal to the other squad's we have control over the building and have established overwatch. The other squads begin assaults and we capture the police station back and the city from enemy forces. The operation is a success. It's a training exercise, but it is exhilarating and empowering. This is how we are to view our fight with the enemy.

In a way, as I am making my way to the end of this short book, I am experiencing some of the same emotions of that day. The feeling of taking charge over the circumstances and the fight. Turning the tides on the enemy and now bringing the fight to his camp, breaching his doors of darkness. This is the power of God to subdue and trample on lions and cobras. The feeling of victory as you stand above your enemy with a sword in hand like David standing over Goliath about to land the final blow.

The Charge

We have been charged to be warriors of God. God is calling us to be victors over our spiritual enemies. He is calling us to be conquerors for Christ in the spirit realm and warriors in His army. He has allowed the enemy to temporarily occupy space and territory, but He is planning the comeback of all times.

He wants to deal with you individually first though. He wants to arm and equip you for the fight ahead. He will train your hands for war and your fingers for battle. (Psalm 144:1)

He will make you resilient both mentally and spiritually. He will search inside your heart and find your resolve.

He is calling you to lead, teach, and preach. In what way I do not know. You will have to be shown this by the Lord. It does not mean you will have to leave your job and become a priest or a missionary.

It simply means in some way, on some scale, be it small or large, you will be called to serve in the war we are facing.

You may not be ready for this yet and that is ok, perhaps I am not ready yet either. Instead, we acknowledge that the Lord has us in a state of preparation and training for the position and mission we are here to accomplish.

Don't discount your presence here on earth during this time. God is not done with you yet. We are constantly in a state of learning and growing in our faith walk up until the day we leave this place to be with the Lord. Don't get complacent, and never assumed you have watched, read, or learned it all. Don't discount the importance of listening to others in the faith community, even if you are in a position of authority or ministry yourself.

And remember, it's important to continue in fellowship because God reveals different parts of the strategy, the mission, and the mystery to different people in different ways. You should always be seeking out the knowledge and heart of God. Pursue peace in your family, your job, and your life, and God's peace will find you.

> Jesus says in John 14:27 says "Peace I leave with you; my peace I give you. I do not give to you as the world gives. Do not let your hearts be troubled and do not be afraid."

This isn't just that awkward moment in church shaking hands, it is a reminder and a call to action. Peace doesn't come from the world, so we should not seek it there. With that said, we are called not to be afraid, because our strength and our salvation lie in Him alone.

God wants you to start fighting back. He wants to send you out onto the battlefield. He will take you as far as He thinks you need to go and as far as you let Him.

You could even end up being spiritual special forces or called to a higher office in his kingdom.

But even if you're not, your job and your mission are just as important to war as anyone else's.

Remember In the military, not everyone is a combat arms soldier. There are numerous support roles, and they are all equally as important to the war. The military functions a lot like the body of Christ with multiple moving pieces and we are all the hands of Jesus.

I pray that this book was enlightening to you, whether you are a military veteran, police officer, or just a warrior in Christ, I hope you will be able to put the concepts to work.

My friend Jesus loves each and every one of you. He wants nothing more than for you to be successful in your missions and victorious in life. Don't waste this time here on earth waiting for the afterlife, your day in paradise will come and you will have earned your reward.

Make use of the time God gave you. Impact and change lives, save souls, and crush serpents.

Be the light God made you to be, rescue those who are lost, and shepherd those who you love. The time is short, and the hour is upon us. You only have one shot at this life. Make the best of it. Now Raise your hand and take the oath, accept Jesus as your personal Lord and savior. Wear your armor every day and keep your spiritual weapons maintained.

If I never have the chance to meet you personally, I'll see you on the other side of eternity.

Love,

Nick

ABOUT THE AUTHOR

Nicholas Anthony is the author of *Invisible Wounds: A Cop's Journey Of Faith Through The Darkness of PTSD*. He began his adult life in uniform serving as both a commissioned Officer in the Army National Guard and as a full-time Police Officer after graduating college at Roger Williams University. Anthony has always had a passion for tactics and training becoming an instructor for his agency in firearms, tactical medicine, and active shooter response. After suffering from Cumulative Post Traumatic Stress Disorder from a career in Law Enforcement, he rediscovered his faith in the scriptures of the Bible and wrote *Invisible Wounds* during some of his darkest moments.

Anthony left his agency in 2021 and moved to Florida where he began to heal and write devotional faith works. After returning to the Law Enforcement world as an Inmate Transport Officer for a county detention center, he began to experience spiritual attacks again. This time, Anthony realized that God was calling him to cast off the "old wineskins" of his former self, leaving the law enforcement world behind for good, for he was being made "new wine" and was being called to serve in the capacity of a new type of warrior, a spiritual one.

Anthony's Substack Publication, *The Ephesians 6:11 Armory* began to draw attention to his Holy Spirit-led devotionals filled with hidden revelations found in the scriptures. It was during this time that Anthony began to realize that God was going to use his former experience, training, and passion for military and law enforcement tactics to show others how to apply warrior concepts to the spiritual fight.

Anthony lives in Florida with his wife Francesca and their two daughters, Aliana and Daniella. Anthony's continued fight on the spiritual battlefield is a living example of the message behind Genesis 50:2 NIV:

> "You intended to harm me, but God intended it for good to accomplish what is now being done, the saving of many lives."

www.ingramcontent.com/pod-product-compliance
Lightning Source LLC
Chambersburg PA
CBHW071248070526
44583CB00017B/2373